T0136845

Springer Theses

Recognizing Outstanding Ph.D. Research

Aims and Scope

The series "Springer Theses" brings together a selection of the very best Ph.D. theses from around the world and across the physical sciences. Nominated and endorsed by two recognized specialists, each published volume has been selected for its scientific excellence and the high impact of its contents for the pertinent field of research. For greater accessibility to non-specialists, the published versions include an extended introduction, as well as a foreword by the student's supervisor explaining the special relevance of the work for the field. As a whole, the series will provide a valuable resource both for newcomers to the research fields described, and for other scientists seeking detailed background information on special questions. Finally, it provides an accredited documentation of the valuable contributions made by today's younger generation of scientists.

Theses are accepted into the series by invited nomination only and must fulfill all of the following criteria

- They must be written in good English.
- The topic should fall within the confines of Chemistry, Physics, Earth Sciences, Engineering and related interdisciplinary fields such as Materials, Nanoscience, Chemical Engineering, Complex Systems and Biophysics.
- The work reported in the thesis must represent a significant scientific advance.
- If the thesis includes previously published material, permission to reproduce this must be gained from the respective copyright holder.
- They must have been examined and passed during the 12 months prior to nomination.
- Each thesis should include a foreword by the supervisor outlining the significance of its content.
- The theses should have a clearly defined structure including an introduction accessible to scientists not expert in that particular field.

More information about this series at http://www.springer.com/series/8790

Juliette Monsel

Quantum Thermodynamics and Optomechanics

Doctoral Thesis accepted by Université Grenoble Alpes, Grenoble, France

Author
Dr. Juliette Monsel
Department of Microtechnology
and Nanoscience
Chalmers University of Technology
Göteborg, Sweden

Supervisor
Dr. Alexia Auffèves
Université Grenoble Alpes
CNRS, Grenoble INP
Institut Néel, Grenoble, France

ISSN 2190-5053 ISSN 2190-5061 (electronic)
Springer Theses
ISBN 978-3-030-54973-2 ISBN 978-3-030-54971-8 (eBook)
https://doi.org/10.1007/978-3-030-54971-8

This Springer imprint is published by the registered company Springer Nature Switzerland AG
The registered company address is: Gewerbestrasse 11, 6330 Cham, Switzerland

Supervisor's Foreword

It is my great pleasure to introduce the Ph.D. manuscript of Dr. Juliette Monsel entitled *Quantum Thermodynamics and Optomechanics*.

Quantum thermodynamics is a recent and exciting field of research lying at the boundary between quantum information, quantum optics and stochastic thermodynamics. It can be seen as a concept factory, that ultimately aims at addressing a wide range of currently open questions. These questions can be fundamental—e.g., what is the nature of time arrow at the quantum scale? Is quantum coherence a new energetic resource?—or show strong practical implications—For instance, what is the energetic cost of counter-acting quantum noise? Will we need a nuclear power plant to build a quantum computer? While the field has attracted lots of interest lately, it has remained mostly theoretical so far. This urges to design realistic proposals to develop experimental quantum thermodynamics. This is, in my opinion, the greatest contribution of the present manuscript.

The research projects that Dr. Monsel has developed during her Ph.D. were all focused on one goal, that is proposing experimental protocols to measure directly work exchanges in the quantum realm. Work is an important concept of thermodynamics, that represents the amount of energy that is received by a system, without its entropy being changed. As such, it quantifies the energetic cost to drive and control a quantum system, possibly against various kinds of noise. Conversely, it also captures the desirable form of energy that can be extracted from free resources like the environmental temperature, by designing so-called heat engines. So far, experiments reporting on work measurement in the quantum realm were based on the ability to monitor the evolution of the quantum thermodynamic system of interest. The amount of work exchanged was thus a reconstructed quantity. In her Ph.D., Dr. Monsel has demonstrated the possibility to measure work "in situ", directly inside the physical entity that drives the thermodynamic system. Two experimental platforms were targeted: Optomechanical devices, and superconducting circuits.

The biggest part of her Ph.D. is devoted to the study of hybrid optomechanical devices as platforms for quantum thermodynamics. Such devices consist of a quantum emitter, parametrically coupled to a mechanical resonator on the one hand,

and to an electromagnetic bath on the other hand. In a quantum thermodynamic protocol, the quantum emitter is the working substance, that exchanges heat with the bath and work with the mechanical resonator. Dr. Monsel has evidenced that the work fluctuations experienced by the quantum emitter are equal to the energetic fluctuations of the mechanical resonator. These energetic fluctuations are shown to verify Jarzynski's equality, a foundational result of stochastic thermodynamics.

These first results bring new and operational tools to overcome a current bottleneck of quantum stochastic thermodynamics. As a matter of fact, the concepts of stochastic work and heat are still the object of discussions. The main problem is that measuring a quantum system can perturb it, and possibly provide energy to it. This purely quantum effect prevents from reconstructing work exchanges in the usual way. According to Dr. Monsel's results, measuring the stochastic work directly inside the battery appears as an experimentally feasible and alternative strategy. It paves the road towards the experimental demonstration of a fluctuation theorem in a quantum open system, a fundamental result that has remained elusive so far.

The second important research path focused on a drastically different platform, namely superconducting circuits. In this framework, the working substance is a superconducting qubit embedded in a circuit. The initially excited qubit releases its energy in the circuit—work being carried by the coherent part of the energy it radiates. Dr. Monsel has evidenced that the amount of work extracted can be tuned, either by stimulating the emission process, or by playing on the amount of quantum coherence initially injected in the qubit's state. This result offers a new evidence of the impact of quantum coherence on thermodynamic performances. It also constitutes a very realistic experimental proposal, that can be implemented in circuit Quantum ElectroDynamics and integrated photonics.

I believe the important corpus of knowledge that Dr. Monsel has gathered along with her Ph.D. will contribute to bring more experimentalists in the field of quantum thermodynamics, and therefore, bring more operational answers to deep open questions.

Quaix en Chartreuse, France
May 2020

Dr. Alexia Auffèves

Acknowledgments

I really appreciated these three years of thesis spent at the Néel Institute. First of all, many thanks to Alexia Auffèves who proposed me this fascinating thesis subject and supervised me. I also thank Maxime Richard and Jean-Philippe Poizat for the discussions we had. My thanks also go to Cyril who was a Ph.D. student there before me and helped me get my thesis off to a good start and to Bogdan who was a pleasant office colleague. My thanks also go to the other Ph.D. students on the team, as well as those who joined us for lunch. I wish good luck to Hippolyte, Laurie and Marco for the continuation of their thesis and to Julian for the end of his. I also wish good luck to Patrice, the newly arrived post-doc. Finally, I would like to thank the NPSC team for their warm welcome and friendly lunches.

Summary

Thermodynamics was developed in the nineteenth century to study steam engines using the cyclical transformations of a working substance to extract heat from thermal baths and convert it into work, possibly stored in a battery. This applied science eventually led to the development of fundamental concepts such as irreversibility. Quantum thermodynamics aims at revisiting these results when the working substances, baths and batteries become quantum systems. Its results are still mainly theoretical. This thesis therefore proposes methods to measure work *in situ*, directly inside the battery, and demonstrate the potential of two platforms to pave the way to the experimental exploration of this fast-growing field.

First, I studied hybrid optomechanical systems which consist of a qubit coupled to the electromagnetic field on the one hand, and to a mechanical resonator on the other hand. The qubit's transition frequency is modulated by the vibrations of the mechanical system that exerts in this way a force on the qubit. The mechanical degree of freedom exchanges work with the qubit and therefore behaves like a dispersive battery, i.e. whose natural frequency is very different from the one of the qubit's transition. Finally, the electromagnetic field plays the role of the bath. I showed that the fluctuations of the mechanical energy are equal to the fluctuations of work, which allows the direct measurement of entropy production. As a result, hybrid optomechanical systems are promising for experimentally testing fluctuation theorems in open quantum systems. In addition, I studied optomechanical energy conversion. I showed that a hybrid optomechanical system can be considered as an autonomous and reversible thermal machine allowing either to cool the mechanical resonator or to build a coherent phonon state starting from thermal noise.

Secondly, I showed that a two-stroke quantum engine extracting work from a single, non-thermal, bath can be made. The qubit is embedded in a one-dimensional waveguide and the battery is the waveguide mode of same frequency as the qubit's transition. Therefore, this is a resonant battery, unlike in the previous case. First, the qubit is coupled to the engineered bath, source of energy and coherence, that makes it relax in an experimentally controllable superposition of energy states. Secondly, the bath is disconnected and work is extracted by driving the qubit with a resonant coherent field. This kind of system, called one-dimensional atom, can be implemented in superconducting or semiconducting circuits. The coherence of the qubit's state improves the performances of this engine both in the regime of classical drive, where a large number of photons is injected in the battery, and in the quantum drive regime of low photon numbers.

This thesis evidences the potential of hybrid optomechanical systems and one-dimensional atoms to explore experimentally on the one hand, irreversibility and fluctuation theorems, and on the other hand, the role of coherence in work extraction.

Contents

Chapter 1
Introduction

Thermodynamics was originally developed in the 19th century to optimize steam engines [1, 2]. In this context, work is defined as useful energy, namely mechanical energy, that can be used to set trains into motion for example. Work exchanges correspond concretely to pushing a piston or lifting a weight. Conversely, heat is energy exchanged with a bath, which is not a mechanical system. Heat may corresponds to energy losses, for instance due to friction. The sum of the work and heat equals the internal energy variation of the studied system, which constitutes the first law of thermodynamics.

At the time, thermodynamics was an engineering science aiming at using the cyclic transformation of a working substance $\mathcal{S}$ to extract heat from baths and convert it into work, possibly stored in a battery, as depicted in Fig. 1.1a. This applied science especially focused on engine efficiency which led to the study of the more fundamental concept of irreversibility [2]. Indeed, the Carnot efficiency, which is the maximum efficiency of an engine operating between two baths, can only be reached when all the transformations in the cycle are reversible [3]. Reversible transformations are always quasi-static, that is very slow. Conversely, irreversibility corresponds to a decrease in efficiency caused by a too fast operation of the thermal machine. Work is always exchanged reversibly with the battery while heat exchanges with the baths are not necessarily reversible.

Besides, irreversibility is quantified by entropy production, the entropy of a system being a measure of its statistical disorder [3]. The second law states that the entropy of an isolated system always increases. Therefore, we can distinguish the past from the future by measuring the entropy, i.e. the entropy production gives us the direction of the arrow of time.

J. Monsel, *Quantum Thermodynamics and Optomechanics*, Springer Theses,
https://doi.org/10.1007/978-3-030-54971-8_1

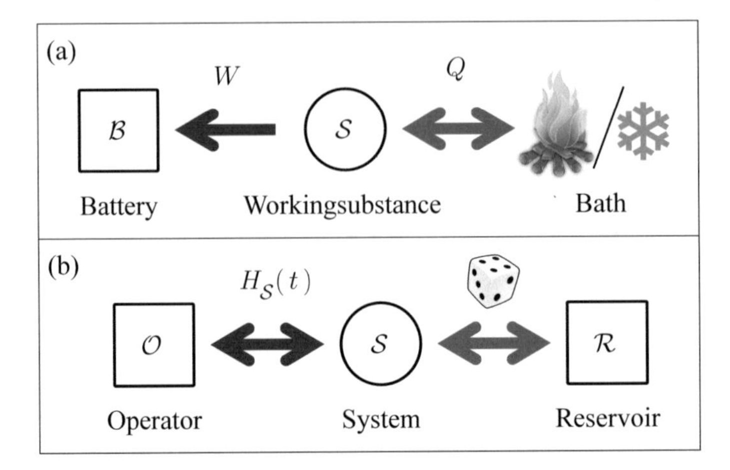

Fig. 1.1 **a** Typical framework in thermodynamics: a working substance $\mathcal{S}$ exchanges work with a battery $\mathcal{B}$ and heat with one or more thermal baths. **b** Typical framework in stochastic thermodynamics: a small system $\mathcal{S}$ is driven by an external operator $\mathcal{O}$, imposing the time-dependent Hamiltonian $H_S(t)$, and randomly disturbed by a thermal reservoir

1.1 Stochastic Thermodynamics

These results from the 19th century give access to average values only, which is sufficient for very large systems, with a number of particles of the order of Avogadro's number, such that energy fluctuations are negligible. However, when the system $\mathcal{S}$ is small, with only a few microstates, fluctuations have an important impact and need to be taken into account. Stochastic thermodynamics [4], developed in the 20th century [5, 6], addresses this new paradigm. This framework, represented in Fig. 1.1b, uses the microscopic description of the system provided by statistical mechanics [7] and models the heat bath as a reservoir $\mathcal{R}$ exerting random forces on the system [8]. Conversely, the external operator $\mathcal{O}$ applies a deterministic force on the system. In this perspective, work is defined as controlled/deterministic energy exchanges while heat is identified with uncontrolled/stochastic energy exchanges [9]. Therefore the operator $\mathcal{O}$ provides work to the system and plays the role of the battery. As the system is randomly disturbed by the bath, it follows a stochastic trajectory in phase space, different for each realization of the same transformation. A well known example is the motion of a Brownian particle in a fluid. It is possible to extend thermodynamic variables at the scale of single trajectories: These are the stochastic heat, work and entropy production [10]. The macroscopic quantities as defined by 19th-century thermodynamics are recovered by averaging over all possible microscopic trajectories.

Stochastic thermodynamics also gives an operational interpretation of the apparent paradox of the observation of irreversibility at the macroscopic scale despite the time-reversibility of the laws of physics at the microscopic scale: Irreversibility comes from the lack of control. After applying some transformation, we can imagine that the operator $\mathcal{O}$ tries to reverse the dynamics of the system to make it go back to its initial state. However, $\mathcal{O}$ can only revert $H_{\mathcal{S}}$ and this is insufficient to make $\mathcal{S}$ follow the time-reversed trajectory in phase space because of the random perturbation caused by the bath. Therefore, an entropy production can be associated with a single trajectory by comparing the probability of this trajectory to occur during the direct transformation to the one of the time-reversed trajectory during the reversed transformation [11].

Another asset of stochastic thermodynamics is that it applies to out-of-equilibrium systems. Especially, many fluctuation theorems linking equilibrium quantities with out-of-equilibrium quantities have been derived [12]. One of the most well known is Jarzynski equality [13] which allows to calculate the variation in equilibrium free energy from the statistics of the work received by a system driven out of equilibrium. Furthermore, in stochastic thermodynamics, the origin of the randomness of the trajectories does not matter, though it was historically thermal. This framework is therefore particularly suitable to investigate thermodynamics at the quantum scale where new sources of randomness appear.

1.2 Quantum Thermodynamics

On the one hand, stochastic thermodynamics started to consider smaller and smaller systems, especially when checking fluctuation theorems: macroscopic oscillators [14], colloidal particles [15–17], single molecules [18], ... Once the systems with quantized energy levels were reached, the question of the extension of fluctuation theorems to quantum systems naturally arose. Another fundamental question raised by the stochastic thermodynamic community is the interpretation of irreversibility in the presence of genuinely quantum sources of randomness, such as quantum measurement [19] and more generally quantum noise [20].

On the other hand, quantum technologies are booming in the wake of the second quantum revolution. The first one occurred at the beginning of the 20th century, describing the rules of physics at the atomic scale and below, discovering the quantization of energy levels and formulating the concept of wave-particle duality [21, 22]. It lead to the understanding of the structure and properties of materials which have allowed the development of electronics and informatics. The second quantum revolution [23] emerged with the tremendous improvement of nanotechnologies which now allows the manipulation and control of single quantum systems. This is about engineering quantum systems to have selected properties and designing quantum circuits with the aim of achieving quantum supremacy [24], i.e. outperforming classical computers. Coherence and entanglement are at the heart of this second revolution. The quantum information community [25] thus naturally became interested in using them as a fuel in quantum engines to outperform classical ones. Another key question that arose is the one of the energy footprints of quantum computations, for instance the cost to create and maintain entanglement or fight against quantum noise.

The recent quantum thermodynamics community [26] results from the merging of scientists from both the stochastic thermodynamics and quantum information communities. The typical framework is the same as in classical stochastic thermodynamics (Fig. 1.1b), except that the working substance, the battery and/or the reservoir are quantum systems, which makes it more difficult to distinguish between work and heat. Therefore, this calls for extensions of the definitions of work, heat and entropy production in this new scenery. There is now a relative consensus about the definition of average work and heat for a quantum system in contact with a thermal bath and driven by a classical operator [27]: work is identified with the energy exchanges induced by the operator, i.e. the work rate corresponds to the variation of $H_{\mathcal{S}}$, while heat is the energy exchanged with the bath, associated with the Lindbladian term $\mathcal{L}[\rho_{\mathcal{S}}]$ in the equation of evolution of the density operator $\rho_{\mathcal{S}}$. As measurements have stochastic outcomes, recent efforts focused on the reconstruction of stochastic thermodynamics with a "quantum dice", i.e. replacing the thermal source of randomness by the quantum randomness caused by measurement [28]. Especially, the definition of a new kind of heat, named "Quantum heat", which is provided by the measurement device, has been introduced recently.

Issues arise when trying to formulate a general definition of work. For instance, when the battery is quantum then the ensemble $\{\mathcal{S} + \mathcal{B}\}$ is described by a time-independent Hamiltonian which makes the definition of work used for driven systems inapplicable. Furthermore, work, unlike internal energy, is not a quantum observable [29], i.e. it cannot be associated with a Hermitian operator. Therefore, work has to be defined operatively, by describing the scheme used to measure it. Several such definitions have been proposed [30, 31], sometimes giving rise to contradictory results [32]. One of the first proposed schemes was to perform two projective energy measurements (TPM), one at the beginning and one at the end of the transformation, defining work as the difference between the two outcomes. This definition is consistent with the classical definition of work [33, 34] but its extension to open quantum systems requires to perform a TPM on the bath as well [35, 36], making this scheme complicated experimentally. Furthermore, the use of projective energy measurements destroys all coherences in the system, preventing the exploration of the role of coherences in thermodynamics.

To sum up, in classical physics, a trajectory is unambiguously defined by the temporal sequence of coordinates of the system in phase space and the way the system is monitored does not alter it. On the contrary, for quantum systems, this monitoring alters the trajectory because measurement disturbs the system and contributes to the energy and entropy exchanges. Therefore alternative strategies to measure energy flows need to be developed. As direct measurements are excluded, there were several proposals to use ancillary systems to measure work [37]. Another proposal is to measure heat exchanges by monitoring the bath and use the first law of thermodynamics to obtain the work [38], the key idea being to engineer the bath so that a photon absorption is associated with a photon emission which can be detected. In this thesis, we propose a different alternative, which is to measure work in situ, directly inside a quantum battery.

1.3 Measuring Work in the Quantum Realm: Two Possible Platforms

In the usual thermodynamic setup, the battery is a classical operator driving the system, so the system's evolution is described by a time-dependent Hamiltonian while the operator is not included in the quantum description of the setup. Therefore, replacing this classical operator by a quantum battery allows a more self-consistent description that takes into account the back-action of the system on the battery. In addition, our proposal in Chaps. 3 and 4 only requires to measure the battery at the beginning and at the end of the thermodynamic transformation which is far easier than a time resolved monitoring of the system and/or the bath. More precisely, we investigate the work exchanges between a qubit, that is a two-level system, and a quantum battery. Two different platforms are considered: optomechanical devices and one-dimensional atoms.

1.3.1 Optomechanical Devices

Hybrid optomechanical systems [39, 40] consist of a qubit parametrically coupled to a nanomechanical resonator. This kind of device can be implemented in various platforms, e.g. superconducting qubits embedded in oscillating membranes [41, 42], nanowires coupled to diamond nitrogen vacancies [43], or to semiconductor quantum dots [44]. The physical origin of the coupling depends on the platform: capacitive coupling, magnetic field gradient and strain-mediated coupling respectively for the cited devices. In any case, the effect of this optomechanical coupling is the same: The motion of the resonator induces a modulation of the qubit's transition frequency. The resonator therefore plays the role of the battery, applying the transformation on the qubit. But, unlike for a classical operator, its energy is small enough to be noticeably impacted by work exchanges, making it possible to measure work directly inside the battery. The mechanical frequency is orders of magnitude smaller than the transition frequency of the qubit, so there is no resonance between the two systems. Therefore, in hybrid optomechanical systems, the battery is dispersive.

The bath is the electromagnetic environment of the qubit, that is a reservoir of photons with a thermal distribution. The qubit may also interact with a laser which can be considered as an additional non-thermal bath.

1.3.2 One-Dimensional Atoms

The second studied platform are the so-called "one-dimensional atoms" [45, 46]. The qubit is an artificial atom embedded in a one-dimensional waveguide. It can be driven by injecting a coherent light field in the waveguide. In the classical limit of large photon numbers, this corresponds to classical Rabi oscillations [47]. The mode of the waveguide of same frequency as the qubit's transition plays the role of the battery, which is resonant unlike in optomechanical systems. Therefore this platform is promising to explore the impact of coherence on thermodynamics, especially on engine efficiency. This kind of device can be implemented in superconducting [48, 49] or semiconducting [50, 51] circuits. The work rate is directly obtained through the difference between the outgoing and incoming photon rates which can be measured with a heterodyne detection setup [52]. In this case, the bath is the environment of the qubit.

These devices can be highly sensitive to single-photon effects [53], such as stimulated emission [54]. Moreover, the qubit's decoherence channels can be monitored, enabling the reconstruction of the stochastic trajectory followed by the qubit [55].

1.4 Outline

This thesis consists of five parts. Chapter 2 presents the theoretical framework of open quantum systems used in the following chapters and summarizes the definitions and results of the thermodynamics of such systems when the battery is a classical operator and the bath is thermal. Chapters 3–5 deal with hybrid optomechanical systems, evidencing that these devices are promising platforms to experimentally explore quantum thermodynamics. Namely, Chap. 3 focuses on the average thermodynamics of such systems, showing that the mechanical resonator acts as a dispersive battery and can be used to directly measure average work exchanges. In Chap. 4, we go one step further and show that the battery's energy fluctuations equal work fluctuations. We then use this result to access entropy production and probe fluctuation theorems. Chapter 5 studies hybrid optomechanical systems as autonomous thermal machines and proves that they can perform optomechanical energy conversion. More precisely, shining a red-detuned laser on the qubit leads to a cooling down of the mechanics in a similar way to evaporative cooling. Conversely, if the laser is blue-detuned, the mechanical motion is amplified and we evidence that a coherent phonon state can be built starting from thermal noise. Finally, Chap. 6 is devoted to a different kind of quantum machines where the battery is resonant with the qubit's transition. We demonstrate that a two-stroke engine, cyclically extracting work from a single non-thermal bath, can be made from a one-dimensional atom. We also evidence that coherence plays a key role in heat-to-work conversion.

References

1. Carnot S (1824) Réflexions Sur La Puissance Motrice Du Feu et Sur Les Machines Propres à Développer Cette Puissance. Bachelier, Paris
2. Clausius R (1867) The mechanical theory of heat: with its applications to the steam- engine and to the physical properties of bodies. J. Van Voorst, London
3. Callen HB (1985) Thermodynamics and an introduction to thermostatistics. Wiley, New York
4. Seifert U (2008) Stochastic thermodynamics: principles and perspectives. Eur Phys J B 64(3–4):423–431. https://doi.org/10.1140/epjb/e2008-00001-9
5. Gibbs JW (1902) Elementary principles in statistical mechanics. Charles Scribner's Sons
6. Boltzmann L (1974) The second law of thermodynamics. In: Boltzmann L, McGuinness B (eds) Theoretical physics and philosophical problems: selected writings. Vienna circle collection. Springer Netherlands, Dordrecht, pp 13–32
7. Landau LD, Lifshitz EM (1980) Statistical physics, vol 5. Course of theoretical physics. Butterworth-Heinemann, Amsterdam
8. Onsager L, Machlup S (1953) Fluctuations and irreversible processes. Phys Rev 91(6):1505–1512. https://doi.org/10.1103/PhysRev.91.1505
9. Sekimoto K (2010) Stochastic energetics. Lecture notes in physics. Springer, Berlin
10. Seifert U (2012) Stochastic thermodynamics, fluctuation theorems and molecular machines. Rep Prog Phys 75(12):126 001. https://doi.org/10.1088/0034-4885/75/12/126001
11. Crooks GE (1999) Entropy production fluctuation theorem and the nonequilibrium work relation for free energy differences. Phys Rev E 60(3):2721–2726. https://doi.org/10.1103/PhysRevE.60.2721
12. Sevick E, Prabhakar R, Williams SR, Searles DJ (2008) Fluctuation theorems. Annu Rev Phys Chem 59(1):603–633. https://doi.org/10.1146/annurev.physchem.58.032806.104555
13. Jarzynski C (1997) Equilibrium free-energy differences from nonequilibrium measurements: a master-equation approach. Phys Rev E 56(5):5018–5035. https://doi.org/10.1103/PhysRevE.56.5018
14. Douarche F, Ciliberto S, Petrosyan A, Rabbiosi I (2005) An experimental test of the Jarzynski equality in a mechanical experiment. Europhys Lett 70(5):593–599. https://doi.org/10.1209/epl/i2005-10024-4
15. Carberry DM, Reid JC, Wang GM, Sevick EM, Searles DJ, Evans DJ (2004) Fluctuations and irreversibility: an experimental demonstration of a second-law- like theorem using a colloidal particle held in an optical trap. Phys Rev Lett 92(14):140 601. https://doi.org/10.1103/PhysRevLett.92.140601
16. Trepagnier EH, Jarzynski C, Ritort F, Crooks GE, Bustamante CJ, Liphardt J (2004) Experimental test of Hatano and Sasa's nonequilibrium steady-state equality. Proc Natl Acad Sci USA 101(42):15 038–15 041. https://doi.org/10.1073/pnas.0406405101,
17. Toyabe S, Jiang H-R, Nakamura T, Murayama Y, Sano M (2007) Experimental test of a new equality: measuring heat dissipation in an optically driven colloidal system. Phys Rev E 75(1):011 122. https://doi.org/10.1103/PhysRevE.75.011122
18. Harris NC, Song Y, Kiang C-H (2007) Experimental free energy surface reconstruction from single-molecule force spectroscopy using Jarzynski's equality. Phys Rev Lett 99(6):068 101. https://doi.org/10.1103/PhysRevLett.99.068101
19. Wiseman HM, Milburn GJ (2010) Quantum measurement and control. Cambridge University Press, Cambridge
20. Gardiner C, Zoller P (2004) Quantum noise: a handbook of Markovian and non-Markovian quantum stochastic methods with applications to quantum optics. Springer series in synergetics. Springer, Berlin
21. Feynman RP, Leighton RB, Sands M (1965) Feynman lectures on physics. Quantum mechanics, vol 3. Addison Wesley, Reading
22. Cohen-Tannoudji C, Diu B, Laloe F (1991) Quantum mechanics, vol 1. Wiley, New York
23. MacFarlane AGJ, Dowling JP, Milburn GJ (2003) Quantum technology: the second quantum revolution. Phil Trans R Soc 361(1809):1655–1674. https://doi.org/10.1098/rsta.2003.1227

24. Oliver WD (2019) Quantum computing takes flight. Nature 574(7779):487–488. https://doi.org/10.1038/d41586-019-03173-4
25. Barnett S (2009) Quantum information. Oxford University Press, Oxford
26. Binder F, Correa LA, Gogolin C, Anders J, Adesso G (eds) (2018) Thermodynamics in the quantum regime: fundamental aspects and new directions. Fundamental theories of physics. Springer, Cham
27. Alicki R (1979) The quantum open system as a model of the heat engine. J Phys A: Math Gen 12(5):L103–L107. https://doi.org/10.1088/0305-4470/12/5/007
28. Elouard C, Herrera-Martí DA, Clusel M, Auffèves A (2017) The role of quantum measurement in stochastic thermodynamics. Npj Quantum Inf 3(1):9. https://doi.org/10.1038/s41534-017-0008-4
29. Talkner P, Lutz E, Hänggi P (2007) Fluctuation theorems: work is not an observable. Phys Rev E 75(5):050 102. https://doi.org/10.1103/PhysRevE.75.050102
30. Talkner P, Hänggi P (2016) Aspects of quantum work. Phys Rev E 93(2):022 131. https://doi.org/10.1103/PhysRevE.93.022131
31. Bäumer E, Lostaglio M, Perarnau-Llobet M, Sampaio R (2018) Fluctuating work in coherent quantum systems: proposals and limitations. In: Binder F, Correa LA, Gogolin C, Anders J, Adesso G (eds) Thermodynamics in the quantum regime. Springer, Cham
32. Engel A, Nolte R (2007) Jarzynski equation for a simple quantum system: comparing two definitions of work. Europhys Lett 79(1):10 003. https://doi.org/10.1209/0295-5075/79/10003
33. Jarzynski C, Quan HT, Rahav S (2015) Quantum-classical correspondence principle for work distributions. Phys Rev X 5(3):031 038. https://doi.org/10.1103/PhysRevX.5.031038
34. Zhu L, Gong Z, Wu B, Quan HT (2016) Quantum-classical correspondence principle for work distributions in a chaotic system. Phys Rev E 93(6):062 108. https://doi.org/10.1103/PhysRevE.93.062108
35. Esposito M, Harbola U, Mukamel S (2009) Nonequilibrium fluctuations, fluctuation theorems, and counting statistics in quantum systems. Rev Mod Phys 81(4):1665–1702. https://doi.org/10.1103/RevModPhys.81.1665
36. Campisi M, Hänggi P, Talkner P (2011) Colloquium: quantum fluctuation relations: foundations and applications. Rev Mod Phys 83(3):771–791. https://doi.org/10.1103/RevModPhys.83.771
37. De Chiara G, Solinas P, Cerisola F, Roncaglia AJ (2018) Ancilla-assisted measurement of quantum work. In: Binder F, Correa LA, Gogolin C, Anders J, Adesso G (eds) Thermodynamics in the quantum regime, vol 195. Fundamental theories of physics. Springer, Cham
38. Elouard C, Bernardes NK, Carvalho ARR, Santos MF, Auffèves A (2017) Probing quantum fluctuation theorems in engineered reservoirs. New J Phys 19(10):103 011. https://doi.org/10.1088/1367-2630/aa7fa2
39. Treutlein P, Genes C, Hammerer K, Poggio M, Rabl P (2014) Hybrid mechanical systems. In: Aspelmeyer M, Kippenberg TJ, Marquardt F (eds) Cavity optomechanics: nano- and micromechanical resonators interacting with light. Quantum science and technology. Springer, Berlin, pp 327–351
40. Bowen WP, Milburn GJ (2016) Quantum optomechanics. CRC Press, Taylor & Francis Group, Boca Raton, FL
41. LaHaye MD, Suh J, Echternach PM, Schwab KC, Roukes ML (2009) Nanomechanical measurements of a superconducting qubit. Nature 459(7249):960–964. https://doi.org/10.1038/nature08093
42. Pirkkalainen J-M, Cho SU, Li J, Paraoanu GS, Hakonen PJ, Sillanpää MA (2013) Hybrid circuit cavity quantum electrodynamics with a micromechanical resonator. Nature 494(7436):211–215. https://doi.org/10.1038/nature11821
43. Arcizet O, Jacques V, Siria A, Poncharal P, Vincent P, Seidelin S (2011) A single nitrogen-vacancy defect coupled to a nanomechanical oscillator. Nat Phys 7(11):879. https://doi.org/10.1038/nphys2070
44. Yeo I, de Assis P-L, Gloppe A, Dupont-Ferrier E, Verlot P, Malik NS, Dupuy E, Claudon J, Gérard J-M, Auffèves A, Nogues G, Seidelin S, Poizat J-P, Arcizet O, Richard M (2014) Strain-mediated coupling in a quantum Dot-Mechanical oscillator hybrid system. Nat Nanotech 9:106–110. https://doi.org/10.1038/nnano.2013.274

45. Turchette Q, Thompson R, Kimble H (1995) One-dimensional atoms. Appl Phys B 60:S1–S10
46. Blais A, Huang R-S, Wallraff A, Girvin SM, Schoelkopf RJ (2004) Cavity quantum electrodynamics for superconducting electrical circuits: an architecture for quantum computation. Phys Rev A 69(6):062 320. https://doi.org/10.1103/PhysRevA.69.062320
47. Cohen-Tannoudji C, Dupont-Roc J, Grynberg G (2004) Atom-photon interactions: basic processes and applications. Physics textbook. Wiley, Weinheim-VCH
48. Hoi I-C, Palomaki T, Lindkvist J, Johansson G, Delsing P, Wilson CM (2012) Generation of nonclassical microwave states using an artificial atom in 1D open space. Phys Rev Lett 108(26):263 601. https://doi.org/10.1103/PhysRevLett.108.263601
49. Eichler C, Lang C, Fink JM, Govenius J, Filipp S, Wallraff A (2012) Observation of entanglement between itinerant microwave photons and a superconducting qubit. Phys Rev Lett 109(24):240 501. https://doi.org/10.1103/PhysRevLett.109.240501
50. Giesz V, Somaschi N, Hornecker G, Grange T, Reznychenko B, De Santis L, Demory J, Gomez C, Sagnes I, Lemaître A, Krebs O, Lanzillotti-Kimura ND, Lanco L, Auffèves A, Senellart P (2016) Coherent manipulation of a solid-state artificial atom with few photons. Nat Commun 7:11 986. https://doi.org/10.1038/ncomms11986
51. Ding D, Appel MH, Javadi A, Zhou X, Löbl MC, Söllner I, Schott R, Papon C, Pregnolato T, Midolo L, Wieck AD, Ludwig A, Warburton RJ, Schröder T, Lodahl P (2019) Coherent optical control of a quantum-dot spin-qubit in a waveguide-based spin-photon interface. Phys Rev Applied 11(3):031 002. https://doi.org/10.1103/PhysRevApplied.11.031002
52. Cottet N, Jezouin S, Bretheau L, Campagne-Ibarcq P, Ficheux Q, Anders J, Auffèves A, Azouit R, Rouchon P, Huard B (2017) Observing a quantum Maxwell demon at work. Proc Natl Acad Sci USA 114(29):7561–7564. https://doi.org/10.1073/pnas.1704827114
53. Wallraff A, Schuster DI, Blais A, Frunzio L, Huang R-S, Majer J, Kumar S, Girvin SM, Schoelkopf RJ (2004) Strong coupling of a single photon to a superconducting qubit using circuit quantum electrodynamics. Nature 431(7005):162–167. https://doi.org/10.1038/nature02851
54. Valente D, Portolan S, Nogues G, Poizat JP, Richard M, Gérard JM, Santos MF, Auffèves A (2012) Monitoring stimulated emission at the single-photon level in one-dimensional atoms. Phys Rev A 85:023 811. https://doi.org/10.1103/PhysRevA.85.023811
55. Ficheux Q, Jezouin S, Leghtas Z, Huard B (2018) Dynamics of a qubit while simultaneously monitoring its relaxation and dephasing. Nat Commun 9(1):1–6. https://doi.org/10.1038/s41467-018-04372-9

Chapter 2
Thermodynamics of Open Quantum Systems

Many situations studied in quantum thermodynamics involve open quantum systems. Indeed, as shown in Fig. 2.1, the typical framework of such systems is very similar to the one of stochastic thermodynamics: a quantum system $\mathcal{S}$ is driven by an external operator $\mathcal{O}$ and weakly coupled to a thermal reservoir $\mathcal{R}$. Before tackling more complex situations, like quantum batteries or non-thermal reservoirs, we need to lay out the key definitions and concepts of quantum thermodynamics in this simpler case where the bath is thermal and the battery is a classical operator.

This chapter therefore summarizes the thermodynamics of a quantum driven system weakly coupled to a single thermal reservoir. First, brief reminders about the theory of open quantum systems are given, introducing the notations. Then, the definitions and key laws of quantum thermodynamics in this context are presented.

2.1 Reminders About Open Quantum Systems

This part aims at providing the few definitions and equations needed to define thermodynamic quantities. The system $\mathcal{S}$ is a possibly driven quantum system interacting with a reservoir $\mathcal{R}$. We further assume that $\mathcal{R}$ is Markovian, i.e. that its correlation time τ_c is negligible compared to the other relevant time scales, and that the coupling is weak, meaning that the influence of $\mathcal{S}$ on $\mathcal{R}$ is small. We discretize time using a time step Δt, chosen to be a lot longer than τ_c but smaller than the characteristic timescale of the system's evolution. Because of our assumptions, the dynamics is Markovian and we can picture the system as interacting with a fresh copy of the reservoir every Δt, like in collisional models or repeated interaction schemes [1, 2], $\mathcal{S}$ and $\mathcal{R}$ being in a product state at the beginning of each time step and the state of $\mathcal{R}$ being reset.

J. Monsel, *Quantum Thermodynamics and Optomechanics*, Springer Theses,
https://doi.org/10.1007/978-3-030-54971-8_2

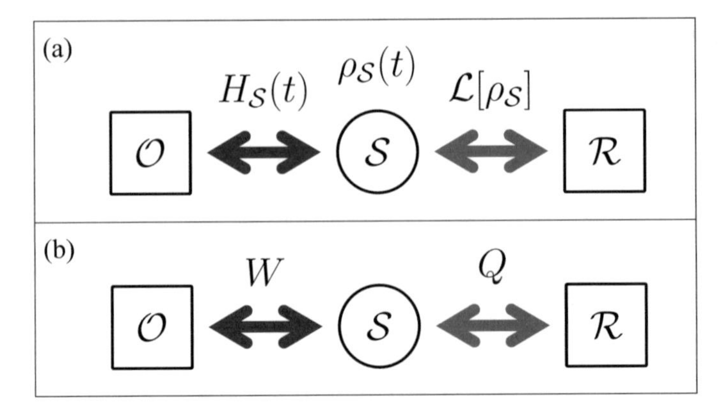

Fig. 2.1 **a** Typical framework in quantum open systems: a system $\mathcal{S}$ is driven by a classical operator $\mathcal{O}$ and coupled to a reservoir $\mathcal{R}$. Due to the external drive, the system's Hamiltonian is time-dependent. The interaction with the reservoir result in a non-unitary evolution, described by the superoperator $\mathcal{L}$ applied on the density operator ρ_S of the system. **b** Typical situation in thermodynamics: a system $\mathcal{S}$ receives work from an operator $\mathcal{O}$

2.1.1 Markovian Master Equation

2.1.1.1 Kraus Sum Representation

$\mathcal{S}$ is not isolated, therefore the evolution of its density operator ρ_S over a time step Δt is not Hamiltonian but given by a dynamical map $\mathcal{M}_{\Delta t}$ such that

$$\rho_S(t + \Delta t) = \mathcal{M}_{\Delta t}[\rho_S(t)]. \tag{2.1}$$

The relationship between the unitary evolution of the closed system $\{\mathcal{S} + \mathcal{R}\}$ and the non-unitary evolution of the subsystem $\mathcal{S}$ is represented in Fig. 2.2. The density operator of $\mathcal{S}$ (resp. $\mathcal{R}$) is obtained by tracing over the degrees of freedom of the other subsystem, i.e. $\rho_{\mathcal{S}/\mathcal{R}} = \text{Tr}_{\mathcal{R}/\mathcal{S}}(\rho_{\text{tot}})$, where ρ_{tot} is the density operator of the whole system. To be physical, $\mathcal{M}_{\Delta t}$ has to transform a density operator into another density operator. Mathematically this translates into $\mathcal{M}_{\Delta t}$ being a convex-linear, completely positive and trace-preserving quantum operation [3, 4]. Any map fulfilling these requirements can be written in the so-called Kraus sum representation [5]

$$\mathcal{M}_{\Delta t}[\rho_S(t)] = \sum_\mu M_\mu(\Delta t)\rho_S(t)M_\mu^\dagger(\Delta t), \tag{2.2}$$

where $\{M_\mu(\Delta t)\}_\mu$ is a collection of operators satisfying the normalization condition

$$\sum_\mu M_\mu^\dagger(\Delta t)M_\mu(\Delta t) = \mathbf{1}. \tag{2.3}$$

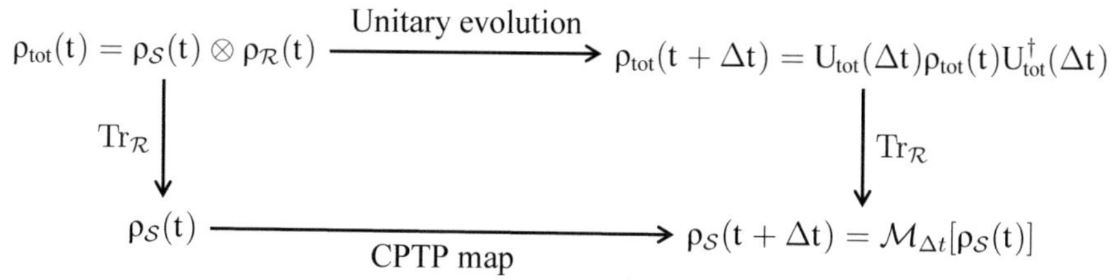

$$
\begin{array}{ccc}
\rho_{\mathrm{tot}}(t) = \rho_{\mathcal{S}}(t) \otimes \rho_{\mathcal{R}}(t) & \xrightarrow{\text{Unitary evolution}} & \rho_{\mathrm{tot}}(t+\Delta t) = U_{\mathrm{tot}}(\Delta t)\rho_{\mathrm{tot}}(t)U_{\mathrm{tot}}^{\dagger}(\Delta t) \\
\downarrow \mathrm{Tr}_{\mathcal{R}} & & \downarrow \mathrm{Tr}_{\mathcal{R}} \\
\rho_{\mathcal{S}}(t) & \xrightarrow[\text{CPTP map}]{} & \rho_{\mathcal{S}}(t+\Delta t) = \mathcal{M}_{\Delta t}[\rho_{\mathcal{S}}(t)]
\end{array}
$$

Fig. 2.2 Schematic representation of the dynamics of the total system and of the subsystem of interest $\mathcal{S}$. The ensemble $\{\mathcal{S} + \mathcal{R}\}$ evolves unitarily under the action of the evolution operator $U_{\mathrm{tot}}(\Delta t)$. The density operator of the subsystem $\mathcal{S}$ is obtained by tracing over the reservoir's degrees of freedom and therefore, its evolution is not unitary in general but described by a completely positive and trace preserving map (CPTP) $\mathcal{M}_{\Delta t}$. The fact that the total system is taken in a product state at time t comes from the assumption of weak coupling [3]

This representation is not unique, but there always exists one with a finite number of operators, not greater than the dimension of the Hilbert space of $\mathcal{S}$ squared.

2.1.1.2 Lindblad Master Equation

Given that the reservoir is weakly coupled to the system, in addition to the Kraus sum representation, the evolution of $\mathcal{S}$ can be given in the form of a Lindblad master equation [3, 4]:

$$\dot{\rho}_{\mathcal{S}}(t) = -\frac{\mathrm{i}}{\hbar}[H_{\mathcal{S}}(t), \rho_{\mathcal{S}}(t)] + \mathcal{L}[\rho_{\mathcal{S}}(t)]. \tag{2.4}$$

This equation is a coarse-grained approximation of the evolution of the density operator, the derivative of the density operator $\rho_{\mathcal{S}}$ being approximated by

$$\dot{\rho_{\mathcal{S}}}(t) = \frac{\rho_{\mathcal{S}}(t+\Delta t) - \rho_{\mathcal{S}}(t)}{\Delta t}. \tag{2.5}$$

The Lindblad superoperator $\mathcal{L}$ reads

$$\mathcal{L}[\rho_{\mathcal{S}}(t)] = \sum_{\mu=1}^{\mathcal{N}} D[L_{\mu}]\rho_{\mathcal{S}}(t), \tag{2.6}$$

and we have used the notation

$$D[X]\rho := X\rho X^{\dagger} - \frac{1}{2}(X^{\dagger}X\rho + \rho X^{\dagger}X). \tag{2.7}$$

The discrete set of operators $\{L_{\mu}\}_{\mu=1}^{\mathcal{N}}$, called jump operators (See Sect. 2.1.2.1), is not unique and is related to the following choice of Kraus sum representation:

$$M_0(\Delta t) = \mathbf{1} - \frac{\mathrm{i}\Delta t}{\hbar} H_{\mathcal{S}}(t) - \frac{\Delta t}{2} \sum_{\mu=1}^{\mathcal{N}} L_\mu^\dagger L_\mu, \tag{2.8}$$

$$M_\mu(\Delta t) = \sqrt{\Delta t} L_\mu, \mu = 1, \ldots, \mathcal{N}. \tag{2.9}$$

The Lindblad master equation (2.4) gives us the average evolution of $\mathcal{S}$. However, to access energy fluctuations, we need to describe the evolution of $\mathcal{S}$ in more details and to be able to define the trajectory followed by $\mathcal{S}$ during the transformation. The next section therefore explains how to unravel a master equation into quantum trajectories.

2.1.2 *Quantum Trajectories*

As the set of operators $\{M_\mu(\Delta t)\}_\mu$ fulfills the normalization condition (2.3), it can be seen as a generalized quantum measurement [6]. The indices μ corresponds to the possible outcomes. Equation (2.2) can therefore be interpreted as the state of the system after such a measurement when the outcome is not read. The Lindblad master equation (2.4) therefore gives the evolution of $\rho_{\mathcal{S}}$ as if the environment was measuring the system every Δt but without reading the outcome.

Conversely, if the measurement outcome is read, then $\rho_{\mathcal{S}}(t + \Delta t)$ is no longer given by Eq. (2.1) but by

$$\rho_{\mathcal{S}}(t + \Delta t) = \frac{M_r \rho_{\mathcal{S}}(t) M_r^\dagger}{\mathrm{Tr}(\rho_{\mathcal{S}}(t) M_r^\dagger M_r)}, \tag{2.10}$$

where r is the measurement outcome. Since this measurement process is stochastic, the dynamics of the system is no longer deterministic but takes the form of a stochastic quantum trajectory $\vec{\Sigma}$ depending on the measurement outcomes. Nevertheless, the same evolution as given by the Lindblad master equation is recovered when averaging over all possible trajectories.

This quantum trajectory picture of the dynamics of the system was originally developed as a numerical simulation tool to overcome computational issues arising when trying to integrate the master equation for large Hilbert spaces [7]. Therefore, the trajectories obtained by this method were seen as virtual paths. Interestingly, with the improvement of nanotechnologies, it is now possible to experimentally keep track of these trajectories [8–11], evidencing their physical relevance.

An unraveling of a Lindblad master equation is the choice of a particular generalized quantum measurement, i.e. of a set of Kraus operators $\{M_\mu(\Delta t)\}_\mu$, that will allow us to gather information on the system without changing its average evolution. In the following, we will focus on the two kinds of unravelings that are used in this thesis: the quantum jump unraveling and the quantum state diffusion. In both cases, we assume that the system $\mathcal{S}$ is always initially prepared in a pure state. Time is discretized using the time step Δt, which corresponds to the interval between two measurements, and we define $t_n = n\Delta t$. Therefore, a trajectory $\vec{\Sigma}$ is a sequence of pure state $\{|\psi_\Sigma(t_n)\rangle\}_{n=0}^{N}$ where N is the total number of measurements performed. Denoting $\{r_\Sigma(t_n)\}_{n=0}^{N-1}$ the measurement record associated with the trajectory $\vec{\Sigma}$, the state of the system at time t_{n+1}, with $n \in \{0, \ldots, N-1\}$, reads

$$|\psi_\Sigma(t_{n+1})\rangle = \frac{M_{r_\Sigma(t_n)}}{\sqrt{p_{r_\Sigma(t_n)}}} |\psi_\Sigma(t_n)\rangle \,, \tag{2.11}$$

where

$$p_{r_\Sigma(t_n)} = \left\langle \psi_\Sigma(t_n) \left| M^\dagger_{r_\Sigma(t_n)} M_{r_\Sigma(t_n)} \right| \psi_\Sigma(t_n) \right\rangle \tag{2.12}$$

is the probability to obtain the outcome $r_\Sigma(t_n)$ when measuring $|\psi_\Sigma(t_n)\rangle$. The density operator $\rho_\mathcal{S}$ as given by the Lindblad master equation (2.4) is then recovered by averaging over the trajectories:

$$\begin{aligned} \rho_\mathcal{S}(t_n) &= \langle |\psi_\Sigma(t_n)\rangle\langle\psi_\Sigma(t_n)| \rangle_{\vec{\Sigma}} \\ &= \sum_{\vec{\Sigma}} P[\vec{\Sigma}]\, |\psi_\Sigma(t_n)\rangle\langle\psi_\Sigma(t_n)| \,. \end{aligned} \tag{2.13}$$

$P[\vec{\Sigma}]$ is the probability that the system follows the stochastic trajectory $\vec{\Sigma}$.

It is important to note that unlike their classical counterparts, quantum trajectories are closely related to the chosen measurement scheme. Indeed, in classical physics, a trajectory is unambiguously defined by the temporal sequence of coordinates of the system in phase space and the way the system is monitored does not alter it. On the contrary, for quantum systems, different choices of measurement schemes result in totally different trajectories.

2.1.2.1 Quantum Jump Unraveling

For this specific unraveling, the chosen generalized quantum measurement is the discrete set of operators $\{M_\mu\}_{\mu=0}^{\mathcal{N}}$ defined by Eqs. (2.8) and (2.9) from the Lindblad master equation. Therefore, at each time step n, the system will evolve in one of two very different ways. If the measurement outcome $r_\Sigma(t_n)$ is 0, the system evolves under the action of the operator M_0. Since M_0 is of order unity, the system's state changes very little during the time Δt and this evolution can be seen as the infinitesimal change in a continuous evolution described by the non-Hermitian Hamiltonian [4]

$$H_{\text{eff}}(t_n) = H_\mathcal{S}(t_n) - \mathrm{i}\frac{\hbar}{2}\sum_{\mu=1}^{\mathcal{N}} L^\dagger_\mu L_\mu. \tag{2.14}$$

On the contrary, if $r_\Sigma(t_n) > 0$, then the system undergoes an important change and "jumps" from state $|\psi_\Sigma(t_n)\rangle$ to the very different state $L_{r_\Sigma(t_n)}\,|\psi_\Sigma(t_n)\rangle$ (up to a normalization), hence the name "jump operators" to designate the operators $\{L_\mu\}_{\mu=1}^{\mathcal{N}}$ in Sect. 2.1.1.2. This unraveling will be used in Chaps. 4 and 5.

To give a concrete example of quantum trajectories, we apply the quantum jump unraveling to the master equation of a non-driven qubit in contact with a reservoir at zero temperature [4]:

$$\dot{\rho}_q(t) = -\frac{i}{\hbar}[\hbar\omega_0 |e\rangle\langle e|, \rho_q(t)] + \gamma D[\sigma_-]\rho_q(t). \tag{2.15}$$

ρ_q is the density operator of the qubit of excited (resp. ground) state $|e\rangle$ (resp. $|g\rangle$) and transition frequency ω_0. We have defined $\sigma_- = |g\rangle\langle e|$. There is a single jump operator $L = \sqrt{\gamma}\sigma_-$ corresponding to the emission of one photon at frequency ω_0. This unraveling corresponds to the replacement of the environment by a single-photon detector. At each time step Δt, either the detector clicks, indicating the emission of a photon, i.e. there was a jump and the qubit ends in the ground state, or there is no click and the qubit evolves under the action of the effective non-Hermitian Hamiltonian $H_{\text{eff}} = \hbar(\omega_0 - i\gamma/2)|e\rangle\langle e|$. These two alternatives for the n-th time step sum up to:

- Click with probability $p = \gamma\Delta t|\langle\psi_\Sigma(t_n)|e\rangle|^2$:

$$|\psi_\Sigma(t_{n+1})\rangle = |g\rangle, \tag{2.16}$$

- No click with probability $1 - p$:

$$|\psi_\Sigma(t_{n+1})\rangle = \frac{\mathbf{1} - (i\omega_0 + \gamma/2)\Delta t\, |e\rangle\langle e|}{\sqrt{1-p}} |\psi_\Sigma(t_n)\rangle. \tag{2.17}$$

The evolution of the population P_e of the excited state can then be reconstructed from the detector's click record and will look like the curves in Fig. 2.3 if we initially prepare the qubit in state $|e\rangle$. Though during a trajectory (solid lines) the qubit is always in one of its energy eigenstates, we recover the exponentially decaying population predicted by the master equation (2.15) by averaging P_e over all possible trajectories (dashed line). This kind of trajectories have been observed experimentally, for instance for a microwave photon in superconducting cavity [8] or a superconducting artificial atom [10].

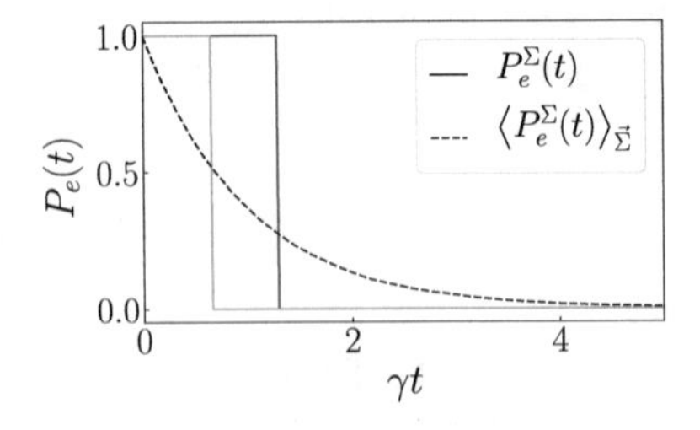

Fig. 2.3 Evolution of the population of the excited state of the qubit P_e in the case where the qubit is initially in state $|e\rangle$. The solid lines are two examples of trajectories $\vec{\Sigma}$ while the dashed line is the evolution of P_e averaged over 10^4 such trajectories

2.1.2.2 Quantum State Diffusion

Quantum state diffusion is another often used kind of unraveling. In this case, the measurement outcome $r_\Sigma(t_n)$ takes continuous values, i.e. there is an infinite number of Kraus operators in the chosen representation. This unraveling corresponds to a situation where the environment is weakly measuring the system, thus the evolution of the system at each time step is always infinitesimal. Equation (2.4) is typically unraveled into [12]

$$\begin{aligned} |\psi_\Sigma(t_{n+1})\rangle = & \left(\mathbf{1} - \frac{\mathrm{i}}{\hbar} H_S(t_n)\right) |\psi_\Sigma(t_n)\rangle \, \Delta t \\ & + \sum_{\mu=1}^{\mathcal{N}} \left(l_\mu^*[\Sigma, t_n] L_\mu - \frac{1}{2} L_\mu^\dagger L_\mu - \frac{1}{2} |l_\mu[\Sigma, t_n]|^2 \right) |\psi_\Sigma(t_n)\rangle \, \Delta t \\ & + \sum_{\mu=1}^{\mathcal{N}} (L_\mu - l_\mu[\Sigma, t_n]) \, |\psi_\Sigma(t_n)\rangle \, \mathrm{d}\xi_\mu, \end{aligned} \tag{2.18}$$

where $l_\mu[\Sigma, t_n] = \langle \psi_\Sigma(t_n) | L_\mu | \psi_\Sigma(t_n) \rangle$. The $\mathrm{d}\xi_\mu$ are $\mathcal{N}$ independent complex Wiener increments, i.e. Gaussian random variables of zero mean and variance Δt. This unraveling will be used in Chap. 5. The typical experimental methods to obtain quantum state diffusion unravelings are homodyne and heterodyne detection schemes [6]. For instance, the trajectories of a superconducting qubit were reconstructed from the heterodyne detection of its fluorescence [11].

2.2 Average Thermodynamics

As previously, we consider a quantum driven system $\mathcal{S}$ in contact with a thermal bath $\mathcal{R}$ at temperature T. The Hamiltonian of the system is therefore time-dependent and the thermodynamic transformation is performed by the external operator that drives the system from $H_\mathcal{S}(t_0)$ to $H_\mathcal{S}(t_\mathrm{f})$. In the rest of this chapter, we assume that the drive preserves the energy eigenstates of the bare system, changing only the energy eigenvalues. This excludes coherent driving like the Rabi oscillations of a two-level system driven by a resonant laser. In this part we will use the results from Sect. 2.1.1 to define the average thermodynamic quantities and assume that the evolution of the density operator $\rho_\mathcal{S}$ is described by the Lindblad master equation (2.4). The system is initially prepared in a mixed state $\rho_\mathcal{S}(t_0)$. We denote $\rho_\mathcal{S}^\infty(H)$ the equilibrium state of the system associated with the time-independent Hamiltonian H, defined by

$$\rho_\mathcal{S}^\infty(H) = \frac{\exp(-H/k_\mathrm{B} T)}{Z(H)}, \tag{2.19}$$

where

$$Z_{\mathcal{S}}(H) = \text{Tr}\,(\exp(-H/k_{\text{B}}T)) \tag{2.20}$$

is the partition function of the system and k_{B} is the Boltzmann constant.

2.2.1 *First Law*

At any time t, the internal energy of the system is defined as the expectation value of the Hamiltonian [13]:

$$\varepsilon_{\mathcal{S}}(t) := \text{Tr}(\rho_{\mathcal{S}}(t)H_{\mathcal{S}}(t)). \tag{2.21}$$

Looking at the system's infinitesimal energy variation over one time step $\text{d}\varepsilon_{\mathcal{S}}(t)$, we get

$$\begin{aligned} \text{d}\varepsilon_{\mathcal{S}}(t) &= \varepsilon_{\mathcal{S}}(t+\Delta t) - \varepsilon_{\mathcal{S}}(t) \\ &= \text{Tr}(\text{d}\rho_{\mathcal{S}}(t)H_{\mathcal{S}}(t)) + \text{Tr}(\rho_{\mathcal{S}}(t)\text{d}H_{\mathcal{S}}(t)). \end{aligned} \tag{2.22}$$

$\text{Tr}(\rho_{\mathcal{S}}(t)\text{d}H_{\mathcal{S}}(t))$ is an energy change induced by the operator, i.e. corresponding to controlled energy exchanges. On the other hand, using Eq. (2.4), we can write

$$\text{Tr}(\text{d}\rho_{\mathcal{S}}(t)H_{\mathcal{S}}(t)) = \text{Tr}(\Delta t\mathcal{L}[\rho_{\mathcal{S}}(t)]H_{\mathcal{S}}(t)) \tag{2.23}$$

and associate this term with uncontrolled energy exchanges with the reservoir. Therefore, in accordance with classical thermodynamics, we define the infinitesimal work and heat received by $\mathcal{S}$ during one time step by [13]

$$\delta W(t) := \text{Tr}(\rho_{\mathcal{S}}(t)\text{d}H_{\mathcal{S}}(t)), \tag{2.24}$$

$$\delta Q(t) := \text{Tr}(\text{d}\rho_{\mathcal{S}}(t)H_{\mathcal{S}}(t)). \tag{2.25}$$

Integrating these equations over time, we obtain the total work and heat received by $\mathcal{S}$ during the transformation:

$$W = \int_{t_0}^{t_{\text{f}}} \text{d}t\text{Tr}(\rho_{\mathcal{S}}(t)\dot{H}_{\mathcal{S}}(t)), \tag{2.26}$$

$$Q = \int_{t_0}^{t_{\text{f}}} \text{d}t\text{Tr}(\dot{\rho}_{\mathcal{S}}H_{\mathcal{S}}(t))). \tag{2.27}$$

The first law of thermodynamics clearly follows from these definitions, both at the infinitesimal and whole transformation scales:

$$d\varepsilon_{\mathcal{S}}(t) = \delta W(t) + \delta Q(t), \tag{2.28}$$
$$\Delta\varepsilon_{\mathcal{S}} = \varepsilon_{\mathcal{S}}(t_{\mathrm{f}}) - \varepsilon_{\mathcal{S}}(t_0) = W + Q. \tag{2.29}$$

2.2.2 Second Law

The Von Neumann entropy of the mixed state $\rho_{\mathcal{S}}$ is defined by

$$S_{\mathrm{VN}}(\rho_{\mathcal{S}}) = -\mathrm{Tr}(\rho_{\mathcal{S}} \log(\rho_{\mathcal{S}})). \tag{2.30}$$

This entropy vanishes for pure states. The entropy change due to the thermodynamic transformation is given by the change in the Von Neumann entropy of the system

$$\Delta S_{\mathcal{S}} = S_{\mathrm{VN}}(\rho_{\mathcal{S}}(t_{\mathrm{f}})) - S_{\mathrm{VN}}(\rho_{\mathcal{S}}(t_0)). \tag{2.31}$$

The irreversibility of the transformation is quantified by the entropy production S_{irr}, defined as the total entropy variation of the ensemble $\{\mathcal{S} + \mathcal{R}\}$ which is isolated. The entropy production obeys the second law of thermodynamics

$$S_{\mathrm{irr}} \geq 0. \tag{2.32}$$

For an isothermal transformation, at temperature T, the entropy change of the system is given by [14]

$$\Delta S_{\mathcal{S}} = S_{\mathrm{irr}} + \frac{Q}{k_{\mathrm{B}}T}, \tag{2.33}$$

like in classical thermodynamics. Therefore, the infinitesimal entropy production during one time step is

$$\begin{aligned} \delta S_{\mathrm{irr}}(t) &= \mathrm{d}S_{\mathcal{S}}(t) - \frac{\delta Q(t)}{k_{\mathrm{B}}T} \\ &= \mathrm{d}S_{\mathcal{S}}(t) - \mathrm{Tr}(\mathrm{d}\rho_{\mathcal{S}}(t)\frac{H_{\mathcal{S}}(t)}{k_{\mathrm{B}}T}), \end{aligned} \tag{2.34}$$

where we have used Eq. (2.25). The Hamiltonian can be expressed as

$$H_{\mathcal{S}}(t) = -k_{\mathrm{B}}T \log(\rho_t^{\infty}) - k_{\mathrm{B}}T \log(Z_t), \tag{2.35}$$

where $\rho_t^{\infty} = \rho^{\infty}(H_{\mathcal{S}}(t))$, given by Eq. (2.19), is the equilibrium state the system would relax into if its Hamiltonian was set to the constant value $H_{\mathcal{S}}(t)$ and $Z_t = Z(H_{\mathcal{S}}(t))$ is the corresponding partition function, given by Eq. (2.20). Therefore, we obtain

$$\delta S_{\mathrm{irr}}(t) = -S(\mathrm{d}\rho_{\mathcal{S}}(t) \| \rho_t^{\infty}), \tag{2.36}$$

where we have defined the relative entropy

$$S(\rho\|\rho') := \mathrm{Tr}(\rho\log(\rho) - \rho\log(\rho')), \tag{2.37}$$

which has the mathematical properties of a distance between ρ and ρ' [15, 16]. The infinitesimal entropy production can thus be interpreted as the variation in the distance between the state of the system and the local equilibrium state. For a quasi-static transformation, the variations of $H_S(t)$ are so slow that the system is always in the local equilibrium state ρ_t^∞ and therefore the entropy production vanishes.

In the case of a simple relaxation, i.e. we have prepared the system in some out-of-equilibrium state $\rho_S(t_0)$ then put it in contact with the reservoir without driving it, H_S and therefore ρ_S^∞ are time-independent. As a consequence, integrating Eq. (2.36), we obtain

$$S_{\mathrm{irr}} = S(\rho_S(t_0)\|\rho_S^\infty), \tag{2.38}$$

assuming that t_{f} is large enough to have $\rho_S(t_f) = \rho_S^\infty$. This result means that the total entropy production is given by the distance to the equilibrium state.

2.3 Stochastic Thermodynamics

To access the fluctuations of the thermodynamic quantities defined in the previous section, we need to go beyond the Lindblad master equation and use the quantum trajectory picture from Sect. 2.1.2. In the following we will therefore assume that an unraveling has been chosen. In practice, this unraveling will be determined by the experimental detection setup (e.g. photo-detection, homodyne detection, ... [6]). During the same transformation as previously, of duration $t_{\mathrm{f}} = N\Delta t$, the system now follows a trajectory $\vec{\Sigma} = \{|\psi_\Sigma(t_n)\rangle\}_{n=0}^N$. The system is initially in the mixed state

$$\rho_S(t_0) = \sum_\lambda p_\lambda\, |\lambda\rangle\langle\lambda|\,, \tag{2.39}$$

therefore $|\psi_\Sigma(t_0)\rangle$ is randomly chosen among the eigenstates $\{|\lambda\rangle\}_\lambda$ of the density operator with probability p_λ [4].

2.3.1 *First Law*

The internal energy of the system along the trajectory $\vec{\Sigma}$ is defined by [17]

$$\varepsilon_S[\vec{\Sigma}, t_n] := \langle\psi_\Sigma(t_n)\,|H_S(t_n)|\,\psi_\Sigma(t_n)\rangle\,. \tag{2.40}$$

This definition is similar to the average one (Eq. (2.21)), except that the density operator $\rho_{\mathcal{S}}(t_n)$ has been replaced by the pure state $|\psi_\Sigma(t_n)\rangle\langle\psi_\Sigma(t_n)|$. The energy variation during the n-th time step reads

$$\begin{aligned}\mathrm{d}\varepsilon_{\mathcal{S}}[\vec{\Sigma}, t_n] &= \varepsilon_{\mathcal{S}}[\vec{\Sigma}, t_{n+1}] - \varepsilon_{\mathcal{S}}[\vec{\Sigma}, t_n] \\ &= \langle\psi_\Sigma(t_{n+1})\,|H_{\mathcal{S}}(t_{n+1})|\,\psi_\Sigma(t_{n+1})\rangle - \langle\psi_\Sigma(t_n)\,|H_{\mathcal{S}}(t_{n+1})|\,\psi_\Sigma(t_n)\rangle \\ &\quad + \langle\psi_\Sigma(t_n)\,|H_{\mathcal{S}}(t_{n+1})|\,\psi_\Sigma(t_n)\rangle - \langle\psi_\Sigma(t_n)\,|H_{\mathcal{S}}(t_n)|\,\psi_\Sigma(t_n)\rangle\,. \end{aligned} \tag{2.41}$$

As previously, we identify work with the energy injected by the drive in the system. Therefore, the work increment during the n-th time step is defined by

$$\delta W[\vec{\Sigma}, t_n] := \langle\psi_\Sigma(t_n)\,|H_{\mathcal{S}}(t_{n+1})|\,\psi_\Sigma(t_n)\rangle - \langle\psi_\Sigma(t_n)\,|H_{\mathcal{S}}(t_n)|\,\psi_\Sigma(t_n)\rangle\,. \tag{2.42}$$

To ensure that the first law is fulfilled for a single time step, the heat increment is defined by

$$\begin{aligned}\delta Q[\vec{\Sigma}, t_n] &:= \mathrm{d}\varepsilon_{\mathcal{S}}[\vec{\Sigma}, t_n] - \delta W[\vec{\Sigma}, t_n] \\ &= \langle\psi_\Sigma(t_{n+1})\,|H_{\mathcal{S}}(t_{n+1})|\,\psi_\Sigma(t_{n+1})\rangle - \langle\psi_\Sigma(t_n)\,|H_{\mathcal{S}}(t_{n+1})|\,\psi_\Sigma(t_n)\rangle\,. \end{aligned} \tag{2.43}$$

These definitions are also similar to the average ones (Eqs. (2.24) and (2.25)) since the work increment is given by the variation of the Hamiltonian while the heat increment is given by the variation of the system's state. The total work and heat received by the system along the trajectory $\vec{\Sigma}$ is then given by summing up the increments:

$$W[\vec{\Sigma}] = \sum_{n=0}^{N-1} \delta W[\vec{\Sigma}, t_n], \tag{2.44}$$

$$Q[\vec{\Sigma}] = \sum_{n=0}^{N-1} \delta Q[\vec{\Sigma}, t_n]. \tag{2.45}$$

Finally, the first law can be written for a complete trajectory:

$$\Delta\varepsilon_{\mathcal{S}}[\vec{\Sigma}] = \varepsilon_{\mathcal{S}}[\vec{\Sigma}, t_N] - \varepsilon_{\mathcal{S}}[\vec{\Sigma}, t_0] = W[\vec{\Sigma}] + Q[\vec{\Sigma}]. \tag{2.46}$$

All these definitions are consistent with the averaged thermodynamic quantities defined in the previous section. Indeed, using Eq. (2.13) we obtain

$$\begin{aligned}\left\langle\!\left\langle\psi_\Sigma(t')\,|H_{\mathcal{S}}(t)|\,\psi_\Sigma(t')\right\rangle\!\right\rangle_{\vec{\Sigma}} &= \sum_{\vec{\Sigma}} P[\vec{\Sigma}]\mathrm{Tr}(H_{\mathcal{S}}(t)\,\big|\psi_\Sigma(t')\big\rangle\!\big\langle\psi_\Sigma(t')\big|) \\ &= \mathrm{Tr}(H_{\mathcal{S}}(t)\rho_{\mathcal{S}}(t')). \end{aligned} \tag{2.47}$$

Then, we can check that $\left\langle \varepsilon_{\mathcal{S}}[\vec{\Sigma}, t_n] \right\rangle_{\vec{\Sigma}} = \varepsilon_{\mathcal{S}}(t_n)$, $\left\langle \delta W[\vec{\Sigma}, t_n] \right\rangle_{\vec{\Sigma}} = \delta W(t_n)$ and $\left\langle \delta Q[\vec{\Sigma}, t_n] \right\rangle_{\vec{\Sigma}} = \delta Q(t_n)$.

2.3.2 *Second Law and Fluctuation Theorems*

2.3.2.1 Entropy

The entropy of the initial state is defined as

$$S_{\mathcal{S}}[\vec{\Sigma}, t_0] := -\log(p_{\rm i}(\psi_\Sigma(t_0))). \tag{2.48}$$

$p_{\rm i}(\psi_\Sigma(t_0))$ is the probability that the system starts in $\psi_\Sigma(t_0)$ and is equal to the eigenvalue p_λ of $\rho_{\mathcal{S}}(t_0)$ associated with the eigenstate $|\lambda\rangle = |\psi_\Sigma(t_0)\rangle$ (See Eq. (2.39)). This definition is consistent with the Von Neumann entropy of the initial state:

$$\left\langle S_{\mathcal{S}}[\vec{\Sigma}, t_0] \right\rangle_{\vec{\Sigma}} = S_{\rm VN}(\rho_{\mathcal{S}}(t_0)). \tag{2.49}$$

Entropy production quantifies irreversibility, i.e. how impossible it is to time-reverse the transformation. Because of the stochastic nature of the quantum trajectories, an entropy production $s_{\rm irr}[\vec{\Sigma}]$ can therefore be associated to the trajectory $\vec{\Sigma}$ by comparing the probability $P[\vec{\Sigma}]$ that the system follows the trajectory $\vec{\Sigma} = \{|\psi_\Sigma(t_n)\rangle\}_{n=0}^N$ during the direct transformation to the probability $\tilde{P}[\overleftarrow{\Sigma}]$ that the system follows the reversed trajectory $\overleftarrow{\Sigma} = \{|\psi_\Sigma(t_n)\rangle\}_{n=N}^0$ during the time-reversed transformation [18]:

$$s_{\rm irr}[\vec{\Sigma}] := \log\left(\frac{P[\vec{\Sigma}]}{\tilde{P}[\overleftarrow{\Sigma}]}\right). \tag{2.50}$$

A more precise definition of the time-reversed transformation will be given for specific cases.

2.3.2.2 Second Law and Absolute Irreversibility

The ratio of the probabilities of the reversed and direct trajectories is given by the exponential of the entropy production:

$$\frac{\tilde{P}[\overleftarrow{\Sigma}]}{P[\vec{\Sigma}]} = \exp(-s_{\rm irr}[\vec{\Sigma}]). \tag{2.51}$$

Therefore, by averaging over all possible direct trajectories, this expression is reduced to

$$\sum_{\overleftarrow{\Sigma}\in\Sigma_{\rm d}} \tilde{P}[\overleftarrow{\Sigma}] = \left\langle \exp(-s_{\rm irr}[\vec{\Sigma}]) \right\rangle_{\vec{\Sigma}}, \tag{2.52}$$

where $\Sigma_{\rm d} = \{\overleftarrow{\Sigma}\,|\,P[\vec{\Sigma}] > 0\}$ is the set of reversed trajectories that have a direct counterpart. This sum is equivalent to the one denoted by $\sum_{\vec{\Sigma}}$ but explicitly ensures that $P[\vec{\Sigma}]$ is positive so that Eq. (2.51) is finite. $\tilde{P}$ is a probability distribution, therefore $0 \leq \sum_{\overleftarrow{\Sigma}\in\Sigma_{\rm d}} \tilde{P}[\overleftarrow{\Sigma}] \leq 1$ and we can define the non negative number σ such that

$$\sum_{\overleftarrow{\Sigma}\in\Sigma_{\rm d}} \tilde{P}[\overleftarrow{\Sigma}] = \exp(-\sigma). \tag{2.53}$$

As a consequence,

$$\left\langle \exp(-s_{\rm irr}[\vec{\Sigma}]) \right\rangle_{\vec{\Sigma}} = \exp(-\sigma), \tag{2.54}$$

which is an example of integral fluctuation theorem (IFT). By convexity of the exponential, we obtain

$$\left\langle \exp(-s_{\rm irr}[\vec{\Sigma}]) \right\rangle_{\vec{\Sigma}} \geq \sigma \geq 0, \tag{2.55}$$

so the definition of entropy production for single trajectory is consistent with the second law.

If all possible trajectories generated by the reversed protocol have a direct counterpart, $\sigma = 0$, so the IFT (2.54) takes the more usual form

$$\left\langle \exp(-s_{\rm irr}[\vec{\Sigma}]) \right\rangle_{\vec{\Sigma}} = 1, \tag{2.56}$$

and we recover the second law exactly. This is typically true when the system is initially in an equilibrium state at the start of both the direct and reversed protocols, which is the case considered in next section. On the contrary, when there exists at least one reversed trajectory without a direct counterpart, then $\sigma > 0$ and the transformation is strictly irreversible. This kind of irreversibility is named absolute irreversibility [19]. It typically arises when the system is not initially prepared in an equilibrium state. For instance, we can consider a single gas particle in a box at temperature T (this is one of the examples given in Ref. [19]). This box is separated in two by a wall and the particle is always initially put in the left part. Then the wall is removed and the particle moves freely in between the two parts due to its thermal motion. For the reversed process, the particle is initially in thermal equilibrium, i.e. with a position randomly chosen with a thermal distribution, then the wall is put back. If the particle is on the left, this reversed trajectory has a direct counterpart, whereas if the particle is on the right, the associated direct trajectory has a zero probability to occur. A more complex case where absolute irreversibility arises is presented in Chap. 4.

2.3.2.3 Jarzynski Equality for a Driven Qubit

This part aims at proving Jarzynski equality [20]

$$\left\langle \exp\left(-\frac{W[\vec{\Sigma}]}{k_B T}\right)\right\rangle_{\vec{\Sigma}} = \exp\left(-\frac{\Delta F}{k_B T}\right) \tag{2.57}$$

in the specific case of a driven qubit in contact with a thermal reservoir. To do so, we will first express the probabilities of the direct and reversed trajectories, then compute the ratio of the two and eventually average it over the trajectories.

We assume that the drive only affects the qubit's transition frequency, namely the qubit's Hamiltonian reads

$$H_q(t) = \hbar\omega(t)\,|e\rangle\langle e|\,. \tag{2.58}$$

The qubit is initially prepared in the equilibrium state $\rho_{t_0}^\infty$ at the start of the transformation. The equilibrium state ρ_t^∞ is given by

$$\rho_t^\infty = p_e^\infty(t)\,|e\rangle\langle e| + p_g^\infty(t)\,|g\rangle\langle g|\,, \tag{2.59}$$

where $p_\epsilon^\infty(t), \epsilon \in \{e, g\}$ is the Boltzmann probability

$$p_\epsilon^\infty(t) = \frac{\exp(-\varepsilon_q(\epsilon)/k_B T)}{Z_t}. \tag{2.60}$$

We have denoted $\varepsilon_q(\epsilon)$ the energy of the state $|\epsilon\rangle$

$$\varepsilon_q(\epsilon) = \hbar\omega(t)\delta_{\epsilon,e}, \tag{2.61}$$

where $\delta_{\epsilon,e}$ is the Kronecker delta, and Z_t the partition function

$$Z_t = 1 + \exp(-\hbar\omega(t)/k_B T). \tag{2.62}$$

Moreover, we assume the evolution of the transition frequency $\omega(t)$ is such that the master equation

$$\dot{\rho}_q = -\frac{i}{\hbar}[H_q(t), \rho_q(t)] + \gamma(t)(\bar{n}(t)+1)D[\sigma_-]\rho_q(t) + \gamma(t)\bar{n}(t)D[\sigma_+]\rho_q(t) \tag{2.63}$$

holds (this is for instance true for an adiabatic driving [21, 22]). Because of the drive, the spontaneous emission rate γ of the qubit and the average number of photons $\bar{n}$ at frequency ω in the bath

$$\bar{n}(t) = \left(\exp\left(\frac{\hbar\omega(t)}{k_B T}\right) - 1\right)^{-1} \tag{2.64}$$

are time-dependent.

We now apply the quantum jump unraveling to this master equation using the following Kraus sum representation:

$$M_0(t_n) = \mathbf{1} - \frac{\mathrm{i}\Delta t}{\hbar} H_{\text{eff}}(t_n), \tag{2.65a}$$

$$M_-(t_n) = \sqrt{\gamma(t_n)\Delta t(\bar{n}(t_n)+1)}\sigma_-, \tag{2.65b}$$

$$M_+(t_n) = \sqrt{\gamma(t_n)\Delta t\bar{n}(t_n)}\sigma_+, \tag{2.65c}$$

with $t_n = n\Delta t$, $\sigma_- = |g\rangle\langle e|$ and $\sigma_+ = |e\rangle\langle g|$. $M_-(t_n)$ and $M_+(t_n)$ respectively correspond to the emission and absorption of one photon of frequency $\omega(t_n)$ by the qubit. $M_0(t_n)$ was obtained from Eq. (2.8) and corresponds to the no-jump evolution, given by the effective Hamiltonian (Eq. (2.14))

$$H_{\text{eff}}(t_n) = \hbar\omega(t_n)\,|e\rangle\langle e| - \frac{\mathrm{i}\hbar\gamma(t_n)}{2}\Big((\bar{n}(t_n)+1)\,|e\rangle\langle e| + \bar{n}(t_n)\,|g\rangle\langle g|\Big). \tag{2.66}$$

The initial state of the qubit $|\epsilon_\Sigma(t_0)\rangle$ for a given trajectory $\vec{\Sigma}$ is drawn among $|e\rangle$ and $|g\rangle$ with probability $p^\infty_{\epsilon_\Sigma(t_0)}(t_0)$. The jumps occur in between energy eigenstates which are also the eigenstates of H_{eff}, therefore at any time t_n, the state of the qubit, denoted $|\epsilon_\Sigma(t_n)\rangle$, is either $|e\rangle$ or $|g\rangle$. The probability of the trajectory $\vec{\Sigma}$ can be expressed as [4]

$$P[\vec{\Sigma}] = p^\infty_{\epsilon_\Sigma(t_0)}(t_0)P[\epsilon_\Sigma(t_1)|\epsilon_\Sigma(t_0)]\ldots P[\epsilon_\Sigma(t_N)|\epsilon_\Sigma(t_{N-1})], \tag{2.67}$$

where

$$P[\epsilon_\Sigma(t_{n+1})|\epsilon_\Sigma(t_n)] = \Big\langle \epsilon_\Sigma(t_n)\Big| M^\dagger_{r_\Sigma(t_n)}(t_n)M_{r_\Sigma(t_n)}(t_n)\Big|\epsilon_\Sigma(t_n)\Big\rangle \tag{2.68}$$

is the probability that the qubit ends in state $|\epsilon_\Sigma(t_{n+1})\rangle$ after the n-th time step knowing that it was in state $|\epsilon_\Sigma(t_n)\rangle$ at time t_n. $r_\Sigma(t_n) = 0, -, +$ denotes the kind of event that took place during the time step (no-jump, emission or absorption).

The time-reversed transformation in this case consists in preparing the system in the equilibrium state $\rho^\infty_{t_N}$, then applying the time-reversed drive $\tilde{H}_{\text{q}}(t) = H_{\text{q}}(t_0 + t_N - t)$ between times t_0 and t_N while the interaction with the bath remains unchanged. This leads to the following time-reversed Kraus operators [17, 23–25]:

$$\tilde{M}_0(t_n) = \mathbf{1} + \frac{\mathrm{i}\Delta t}{\hbar} H^\dagger_{\text{eff}}(t_n), \tag{2.69a}$$

$$\tilde{M}_-(t_n) = M_+(t_n), \tag{2.69b}$$

$$\tilde{M}_+(t_n) = M_-(t_n), \tag{2.69c}$$

Therefore, the probability of the reversed trajectory $\overleftarrow{\Sigma}$ can be expressed as

$$\tilde{P}[\overleftarrow{\Sigma}] = p^\infty_{\epsilon_\Sigma(t_N)}(t_N)\tilde{P}[\epsilon_\Sigma(t_{N-1})|\epsilon_\Sigma(t_N)]\ldots\tilde{P}[\epsilon_\Sigma(t_0)|\epsilon_\Sigma(t_1)], \tag{2.70}$$

where

$$\tilde{P}[\epsilon_\Sigma(t_n)|\epsilon_\Sigma(t_{n+1})] = \left\langle \epsilon_\Sigma(t_{n+1}) \left| \tilde{M}^\dagger_{r_\Sigma(t_n)}(t_n)\tilde{M}_{r_\Sigma(t_n)}(t_n) \right| \epsilon_\Sigma(t_{n+1}) \right\rangle \tag{2.71}$$

is the reversed transition probability. Using Eqs. (2.67) and (2.70), the expression of the ratio of the trajectories probabilities reads

$$\frac{\tilde{P}[\overleftarrow{\Sigma}]}{P[\vec{\Sigma}]} = \frac{p^\infty_{\epsilon_\Sigma(t_N)}(t_N)}{p^\infty_{\epsilon_\Sigma(t_0)}(t_0)} \prod_{n=0}^{N-1} \frac{\tilde{P}[\epsilon_\Sigma(t_n)|\epsilon_\Sigma(t_{n+1})]}{P[\epsilon_\Sigma(t_{n+1})|\epsilon_\Sigma(t_n)]}. \tag{2.72}$$

From the expressions of the transition probabilities (2.68), (2.71) and of the Kraus operators (2.65), (2.69), we obtain

$$\frac{\tilde{P}[\epsilon_\Sigma(t_n)|\epsilon_\Sigma(t_{n+1})]}{P[\epsilon_\Sigma(t_{n+1})|\epsilon_\Sigma(t_n)]} = \exp\left(\frac{\delta Q[\vec{\Sigma}, t_n]}{k_B T}\right), \tag{2.73}$$

where $\delta Q[\vec{\Sigma}, t_n]$ is the heat received by the qubit during the time step, as defined by (2.43). As for the ratio of the initial probabilities, it gives, using (2.60),

$$\frac{p^\infty_{\epsilon_\Sigma(t_N)}(t_N)}{p^\infty_{\epsilon_\Sigma(t_0)}(t_0)} = \exp\left(\frac{\varepsilon_q(\epsilon_\Sigma(t_0)) - \varepsilon_q(\epsilon_\Sigma(t_N))}{k_B T}\right)\frac{Z_{t_0}}{Z_{t_N}}. \tag{2.74}$$

Then, Eq. (2.72) becomes, using the first law (2.46),

$$\begin{aligned}\frac{\tilde{P}[\overleftarrow{\Sigma}]}{P[\vec{\Sigma}]} &= \exp\left(\frac{-\Delta\varepsilon_q[\vec{\Sigma}] + \Delta F + Q[\vec{\Sigma}]}{k_B T}\right)\\ &= \exp\left(\frac{-W[\vec{\Sigma}] + \Delta F}{k_B T}\right),\end{aligned} \tag{2.75}$$

where ΔF denotes the equilibrium free energy variation defined by

$$\Delta F = k_B T \log\left(\frac{Z_{t_0}}{Z_{t_N}}\right). \tag{2.76}$$

Therefore, using Eq. (2.50), the entropy production reads

$$s_{\mathrm{irr}}[\vec{\Sigma}] = \frac{1}{k_B T}(W[\vec{\Sigma}] - \Delta F). \tag{2.77}$$

Any reversed trajectory is of the form $\overleftarrow{\Sigma} = \{|\epsilon_\Sigma(t_n)\rangle\}_{n=N}^0$, with $\epsilon_\Sigma(t_n) \in \{e, g\}$, and corresponds to the direct trajectory $\vec{\Sigma} = \{|\epsilon_\Sigma(t_n)\rangle\}_{n=0}^N$ which clearly has a non-zero probability to occur, given the expression of the transition probabilities (2.68) and the prepared initial state $\rho_{t_0}^{\infty}$. Therefore, the IFT (2.56) holds and finally, injecting the expression of the entropy production (2.77), we obtain Jarzynski equality (2.57).

2.4 Summary

In this chapter, we have recapitulated the key results of the theory of open quantum systems that will be used in this dissertation: the description of the average evolution of a system in contact with a Markovian reservoir by a Lindblad master equation and the quantum trajectory picture giving access to single realizations. Then we have given the definitions of the main thermodynamic quantities for a quantum system driven by a classical operator and in contact with a thermal bath.

First, we have defined the average internal energy of the system and used the Lindblad master equation to identify the average heat and work exchanged during the transformation. The first law of thermodynamics naturally follows from these definitions. We have defined the average entropy of the system and given, in the specific case of an isothermal transformation, the expression of the entropy production that quantifies irreversibility. For a simple relaxation toward equilibrium, this entropy production can be interpreted as the distance between the initial state of the system and its equilibrium state.

Secondly, we used the quantum trajectory picture to apply the results of stochastic thermodynamics to a quantum system. We defined heat, work and entropy production for a single trajectory. We showed that the average thermodynamic quantities are recovered by averaging the stochastic ones over the trajectories. Finally, we derived a generic integral fluctuation theorem for the entropy production and applied it to the specific case of a driven qubit to obtain Jarzynski equality.

References

1. Rau J (1963) Relaxation phenomena in spin and harmonic oscillator systems. Phys Rev 129(4):1880–1888. https://doi.org/10.1103/PhysRev.129.1880
2. Ziman M, Bužek V (2005) All (qubit) decoherences: Complete characterization and physical implementation. Phys Rev A 72(2):022 110. https://doi.org/10.1103/PhysRevA.72.022110
3. Breuer H-P, Petruccione F (2002) The theory of open quantum systems. Oxford University Press, New York
4. Haroche S, Raimond J-M (2006) Exploring the quantum: atoms, cavities, and photons. Oxford University Press, Oxford
5. Kraus K (1983) States, effects, and operations: fundamental notions of quantum theory. Lecture notes in physics. Springer, Berlin
6. Wiseman HM, Milburn GJ (2010) Quantum measurement and control. Cambridge University Press, Cambridge
7. Mølmer K, Castin Y, Dalibard J (1993) Monte Carlo wave-function method in quantum optics. J Opt Soc Am B 10(3):524–538. https://doi.org/10.1364/JOSAB.10.000524

8. Gleyzes S, Kuhr S, Guerlin C, Bernu J, Deléglise S, Busk Hoff U, Brune M, Raimond J-M, Haroche S (2007) Quantum jumps of light recording the birth and death of a photon in a cavity. Nature 446(7133):297–300. https://doi.org/10.1038/nature05589
9. Murch KW, Weber SJ, Macklin C, Siddiqi I (2013) Observing single quantum trajectories of a superconducting quantum bit. Nature 502(7470):211–214. https://doi.org/10.1038/nature12539
10. Vijay R, Slichter DH, Siddiqi I (2011) Observation of quantum jumps in a superconducting artificial atom. Phys Rev Lett 106(11):110 502. https://doi.org/10.1103/PhysRevLett.106.110502
11. Campagne-Ibarcq P, Six P, Bretheau L, Sarlette A, Mirrahimi M, Rouchon P, Huard B (2016) Observing quantum state diffusion by heterodyne detection of fluorescence. Phys Rev X 6(1):011 002. https://doi.org/10.1103/PhysRevX.6.011002
12. Gisin N, Percival IC (1992) The quantum-state diffusion model applied to open systems. J Phys A: Math Gen 25(21):5677. https://doi.org/10.1088/0305-4470/25/21/023
13. Alicki R (1979) The quantum open system as a model of the heat engine. J Phys A: Math Gen 12(5):L103–L107. https://doi.org/10.1088/0305-4470/12/5/007
14. Deffner S, Lutz E (2011) Nonequilibrium entropy production for open quantum systems. Phys Rev Lett 107(14):140 404. https://doi.org/10.1103/PhysRevLett.107.140404
15. Wehrl A (1978) General properties of entropy. Rev Mod Phys 50(2):221–260. https://doi.org/10.1103/RevModPhys.50.221
16. Vedral V (2002) The role of relative entropy in quantum information theory. Rev Mod Phys 74(1):197–234. https://doi.org/10.1103/RevModPhys.74.197
17. Elouard C, Herrera-Martí DA, Clusel M, Auffèves A (2017) The role of quantum measurement in stochastic thermodynamics. Npj Quantum Inf 3(1):9. https://doi.org/10.1038/s41534-017-0008-4
18. Seifert U (2005) Entropy production along a stochastic trajectory and an integral fluctuation theorem. Phys Rev Lett 95(4):040 602. https://doi.org/10.1103/PhysRevLett.95.040602
19. Murashita Y, Funo K, Ueda M (2014) Nonequilibrium equalities in absolutely irreversible processes. Phys Rev E 90(4):042 110. https://doi.org/10.1103/PhysRevE.90.042110
20. Jarzynski C (1997) Equilibrium free-energy differences from nonequilibrium measurements: a master-equation approach. Phys Rev E 56(5):5018–5035. https://doi.org/10.1103/PhysRevE.56.5018
21. Albash T, Boixo S, Lidar DA, Zanardi P (2012) Quantum adiabatic Markovian master equations. New J Phys 14(12):123 016. https://doi.org/10.1088/1367-2630/14/12/123016
22. Dann R, Levy A, Kosloff R (2018) Time-dependent Markovian quantum master equation. Phys Rev A 98(5):052 129. https://doi.org/10.1103/PhysRevA.98.052129
23. Crooks GE (2008) Quantum operation time reversal. Phys Rev A 77(3):034 101. https://doi.org/10.1103/PhysRevA.77.034101
24. Manzano G, Horowitz JM, Parrondo JMR (2018) Quantum fluctuation theorems for arbitrary environments: adiabatic and nonadiabatic entropy production. Phys Rev X 8(3):031 037. https://doi.org/10.1103/PhysRevX.8.031037
25. Manikandan SK, Jordan AN (2019) Time reversal symmetry of generalized quantum measurements with past and future boundary conditions. Quantum Stud: Math Found 6:241–268. https://doi.org/10.1007/s40509-019-00182-w

Chapter 3
Average Thermodynamics of Hybrid Optomechanical Systems

Optomechanical coupling was first achieved in optical cavities with one moving-end mirror coupled to a mechanical oscillator (MO) [1, 2] (See Fig. 3.1a). These devices have paved the way for many applications including sensing [3, 4], cooling the MO down close to its ground state [5–7] and preparing the MO in quantum states [8, 9]. Besides, some features of phonon lasing were observed [10, 11] and there were proposals to make phonon lasers using cavity optomechanics [12–14].

More recently, hybrid optomechanical systems, in which the cavity is replaced by a qubit, have been developed (See Fig. 3.1b). Unlike in cavity optomechanics, these devices are non-linear because the qubit saturates at one excitation. The mechanical motion modulates the qubit's transition frequency which makes the MO play the role of the battery [15]. Therefore these devices are particularly promising test-beds for the thermodynamics of quantum systems, as shown in this chapter and the next two. Physical implementations of such devices have been realized on various platforms [16]. For instance:

- A superconducting qubit, based on Josephson junctions, is capacitively coupled to a nanomechanical resonator. The mechanical motion modulates the capacitance which in turn changes the qubit's frequency [17, 18].
- The qubit is a nitrogen vacancy center hosted inside a diamond nanocrystal. The nanocrystal is placed at the extremity of a nanowire. The optomechanical coupling is then achieved with a magnetic field gradient that affects the qubit's frequency depending on the nanowire's position by Zeeman effect [19].
- The qubit is a semiconductor quantum dot situated at the bottom of a conical shaped nanowire whose top can oscillate [20]. The qubit is not centered in the nanowire so that the mechanical strain applied on it varies with the position of the tip of the nanowire (See Fig. 3.1c).

Several experimental implementations [18–20] are close to reaching the ultra-strong coupling regime where a single photon emission or absorption by the qubit has

J. Monsel, *Quantum Thermodynamics and Optomechanics*, Springer Theses,
https://doi.org/10.1007/978-3-030-54971-8_3

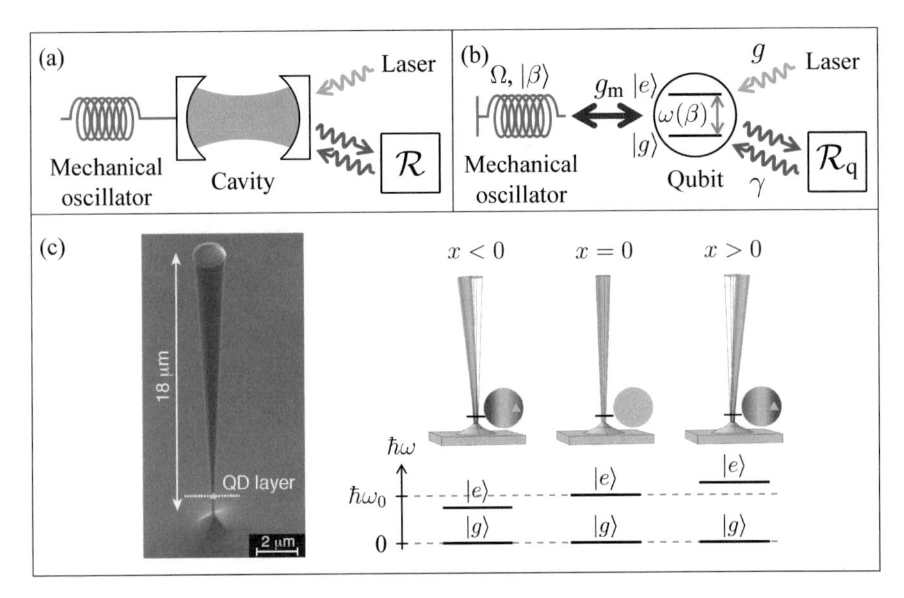

Fig. 3.1 Optomechanical systems. **a** Cavity optomechanics: the light in an optical cavity is coupled to a MO via a moving-end mirror. **b** Hybrid optomechanical system under study: a qubit is dispersively coupled to a nanomechanical oscillator. The qubit can be driven by a laser and is also coupled to an electromagnetic reservoir $\mathcal{R}_q$ of temperature T. **c** Experimental realization of a hybrid optomechanical system consisting of a quantum dot (qubit) embedded in a nanowire (MO). In this device, the optomechanical coupling is mediated by the mechanical strain field around the quantum dot location. Image from the Nanophysics and Semiconductors team of the Institut Néel – CNRS/Université Grenoble Alpes, and the Institute for Nanoscience and Cryogenics of the CEA (Grenoble, France)

a noticeable impact on the mechanics. This regime is the one that will be considered in this thesis.

In this chapter, we apply the theory and definitions from Chap. 2 to a hybrid optomechanical system. First, we present the microscopic model of such systems and derive the master equation for the optomechanical density operator when the qubit is in contact with a thermal reservoir. Then, we study the thermodynamics of the qubit, demonstrating that the mechanical oscillator plays the role of a battery and that average work exchanges can be directly obtained by measuring the oscillator. This last result is particularly interesting since it offers an alternative to the direct measurement of the system to measure work in quantum thermodynamics.

3.1 Dynamics: Master Equation for the Hybrid Optomechanical System

3.1.1 Microscopic Model

3.1.1.1 Hybrid Optomechanical System

We consider a hybrid optomechanical system which consists of a qubit of bare transition frequency ω_0 coupled to a MO of frequency Ω, as depicted in Fig. 3.1b. The Hamiltonian describing such systems reads

$$H_{\mathrm{qm}} = H_{\mathrm{q}} + H_{\mathrm{m}} + V_{\mathrm{qm}} \tag{3.1}$$

where

$$H_{\mathrm{q}} = \hbar\omega_0 \, |e\rangle\langle e| \otimes \mathbf{1}_{\mathrm{m}}, \tag{3.2}$$

$$H_{\mathrm{m}} = \mathbf{1}_{\mathrm{q}} \otimes \hbar\Omega b^\dagger b, \tag{3.3}$$

are the free Hamiltonians of the qubit and MO respectively. We have denoted $|e\rangle$ (resp. $|g\rangle$) the excited (resp. ground) state of the qubit, b the phonon annihilation operator and $\mathbf{1}_{\mathrm{q}}$ (resp. $\mathbf{1}_{\mathrm{m}}$) the identity on the Hilbert space of the qubit (resp. MO). The coupling Hamiltonian is [16]

$$V_{\mathrm{qm}} = \hbar g_{\mathrm{m}} \, |e\rangle\langle e| \, (b^\dagger + b) \tag{3.4}$$

where g_{m} is the optomechanical coupling strength. As a consequence, the qubit's effective transition frequency $\hat{\omega}$ depends on the mechanical position $\hat{x} = x_{\mathrm{zpf}}(b^\dagger + b)$:

$$\hat{\omega} = \omega_0 + g_{\mathrm{m}} \frac{\hat{x}}{x_{\mathrm{zpf}}}, \tag{3.5}$$

where x_{zpf} is the zero-point position fluctuation of the MO. We will denote ρ_{qm} the density operator of the hybrid optomechanical system and $\rho_{\mathrm{q}} = \mathrm{Tr}_{\mathrm{m}}(\rho_{\mathrm{qm}})$ (resp. $\rho_{\mathrm{m}} = \mathrm{Tr}_{\mathrm{q}}(\rho_{\mathrm{qm}})$) the density operator of the qubit (resp. MO).

The Hamiltonian (3.1) can be rewritten

$$H_{\mathrm{qm}} = |g\rangle\langle g| \otimes H_{\mathrm{m}}^{g} + |e\rangle\langle e| \otimes H_{\mathrm{m}}^{e} \tag{3.6}$$

where

$$H_{\mathrm{m}}^{g} = \hbar\Omega b^\dagger b, \tag{3.7a}$$

$$H_{\mathrm{m}}^{e} = \hbar\Omega B^\dagger B + \hbar\left(\omega_0 - \frac{g_{\mathrm{m}}^2}{\Omega}\right)\mathbf{1}_{\mathrm{m}}, \tag{3.7b}$$

and we have defined

$$B = b + \frac{g_{\rm m}}{\Omega}\mathbf{1}_{\rm m} = \hat{D}\left(-\frac{g_{\rm m}}{\Omega}\right) b \hat{D}\left(\frac{g_{\rm m}}{\Omega}\right). \tag{3.8}$$

$\hat{D}$ is the displacement operator, defined by

$$\hat{D}(\alpha) = \exp(\alpha b^\dagger - \alpha^* b), \alpha \in \mathbb{C}. \tag{3.9}$$

Interestingly, we can see from Eq. (3.6) that if the qubit is in either state $|e\rangle$ or state $|g\rangle$, then the MO evolves like a free harmonic oscillator. However, when the qubit is in the excited state, the rest position of the MO is shifted by $-2x_{\rm zpf}g_{\rm m}/\Omega$. The energy eigenbasis of $H_{\rm qm}$ is therefore $\{|g\rangle \otimes |n\rangle, |e\rangle \otimes |n\rangle_B\}_{n\geq 0}$, where $|n\rangle$ is the n-phonon Fock state and $|n\rangle_B$ is the displaced n-phonon state, $|n\rangle_B = \hat{D}\left(-\frac{g_{\rm m}}{\Omega}\right)|n\rangle$. $|g\rangle \otimes |n\rangle$ is associated with the energy $\hbar\Omega n$ and $|e\rangle \otimes |n\rangle_B$ with $\hbar(\Omega n + \omega_0 - g_{\rm m}^2/\Omega)$.

Coherent states of the MO are particularly interesting because they remain coherent states under the action of both $H_{\rm m}^e$ and $H_{\rm m}^g$. A coherent state $|\beta\rangle$, with $\beta \in \mathbb{C}$, is defined by

$$|\beta\rangle = \hat{D}(\beta)|0\rangle, \tag{3.10}$$

with $|0\rangle$ the vacuum state, and is an eigenstate of the phonon annihilation operator:

$$b|\beta\rangle = \beta|\beta\rangle. \tag{3.11}$$

If the qubit is in state $|\epsilon\rangle$, with $\epsilon \in \{e, g\}$, and the MO is in the coherent state $|\beta_0\rangle$ at time $t = 0$, then the state of the MO at time t reads

$$\beta_\epsilon(t) = \left(\beta_0 + \frac{g_{\rm m}}{\Omega}\delta_{\epsilon,e}\right)\exp(-{\rm i}\Omega t) - \frac{g_{\rm m}}{\Omega}\delta_{\epsilon,e}. \tag{3.12}$$

3.1.1.2 Qubit's Bath

As shown in Fig. 3.1b, the qubit can be driven by a laser, but this situation will be considered in Chap. 5 only. The qubit also interacts with an electromagnetic reservoir $\mathcal{R}_{\rm q}$, i.e. a photon reservoir in a thermal state. The Hamiltonian describing the bath reads

$$H_{\mathcal{R}_{\rm q}} = \sum_k \hbar\omega_k a_k^\dagger a_k, \tag{3.13}$$

where a_k is the annihilation operator of the k-th electromagnetic mode of frequency ω_k. $\mathcal{R}_{\rm q}$ is in a thermal state, therefore, it contains on average $\bar{n}_\omega$ photons at any frequency ω, with

$$\bar{n}_\omega = \left(\exp\left(\frac{\hbar\omega}{k_{\rm B}T}\right) - 1\right)^{-1} \tag{3.14}$$

and T the temperature of the reservoir.

The coupling Hamiltonian between the qubit and the bath, in the rotating wave approximation, equals

$$V = \sum_{l=\pm} R_l \otimes \sigma_l \tag{3.15}$$

where $R_+ = \sum_k \hbar g_k a_k$, $R_- = R_+^\dagger$ and g_k is the coupling strength between the qubit and the k-th mode of the reservoir. We denote

$$\gamma_\omega = \sum_k g_k^2 \delta(\omega - \omega_k), \tag{3.16}$$

in particular, $\gamma = \gamma_{\omega_0}$ is the spontaneous emission rate of the bare qubit in contact with the bath.

3.1.2 Studied Regime

In the remaining of this chapter and in Chaps. 4 and 5, we will consider the following regime:

- Dispersive coupling: The qubit and the MO are far from resonance, i.e.

$$\omega_0 \gg \Omega. \tag{3.17}$$

- Ultra-strong coupling: A single photon emission or absorption by the qubit induces a shift in the mechanical rest position larger than the zero-point position fluctuations of the MO, i.e.

$$g_\mathrm{m} \gtrsim \Omega. \tag{3.18}$$

- $\mathcal{R}_\mathrm{q}$ is a Markovian thermal bath of temperature T and therefore,

$$\omega_0 \gg \gamma \text{ and } \tau_\mathrm{c} \ll \gamma^{-1}, g_\mathrm{m}^{-1}, \Omega^{-1}. \tag{3.19}$$

  where τ_c is the correlation time of the bath.

These requirements are met by the experimental implementations cited in the introduction of this chapter, except the one from Ref. [17] (See Table 3.1).

In experimental devices, the MO is also coupled to a phononic thermal reservoir $\mathcal{R}_\mathrm{m}$, giving rise to a mechanical damping rate Γ. However, state-of-the art experimental implementations reach very large quality factors (See Table 3.1) so we can safely neglect the influence of $\mathcal{R}_\mathrm{m}$ on the time scales considered in this chapter and in Chap. 4. This thermal reservoir will be taken into account in Chap. 5.

Table 3.1 Parameters for state-of-the art implementations of hybrid optomechanical systems (Table adapted from [21])

Variable \ Platform			[17]	[18]	[19]	[20]
MO	Frequency	$\Omega/2\pi$	58 MHz	72 MHz	1 MHz	530 kHz
	Quality factor	Ω/Γ	$3 \cdot 10^4$–6.10^4	5500	$>10^4$	$3 \cdot 10^3$
	Zero-point fluctuations	x_{zpf}	13 fm	4 fm	0.7 pm	11 fm
Qubit	Frequency	$\omega_0/2\pi$	~10 GHz	~1 GHz	18 GHz	333 THz
	Spontaneous emission rate	$\gamma/2\pi$	170–800 MHz	3 MHz	7 MHz	157 GHz
Optomechanical coupling strength		g_{m}/Ω	$5 \cdot 10^{-3}$–$40 \cdot 10^{-3}$	0.063–0.35	0.1	0.849

3.1.3 Microscopic Derivation of the Master Equation

We now derive the master equation describing the evolution of the optomechanical system when the qubit is in contact with the reservoir $\mathcal{R}_{\text{q}}$. The total Hamiltonian describing this situation reads

$$H_{\text{tot}} = H_{\text{qm}} + H_{\mathcal{R}_{\text{q}}} + V. \tag{3.20}$$

Given the considered regime, we can define a coarse-graining time step Δt, fulfilling

$$\tau_{\text{c}}, \omega_0^{-1} \ll \Delta t \ll \gamma^{-1}, \Omega^{-1}, g_{\text{m}}^{-1}. \tag{3.21}$$

We assume that the optomechanical system and the bath are initially uncorrelated, therefore the density operator of the total system is of the form

$$\rho_{\text{tot}}(0) = \rho_{\text{qm}}(0) \otimes \rho_{\mathcal{R}_{\text{q}}}(0). \tag{3.22}$$

The hybrid optomechanical system is itself initially prepared in a factorized state

$$\rho_{\text{qm}}(0) = \rho_{\text{q}}(0) \otimes |\beta_0\rangle\langle\beta_0| \tag{3.23}$$

where $\rho_{\text{q}}(0)$ is diagonal in the bare qubit energy basis $\{|e\rangle, |g\rangle\}$ and $|\beta_0\rangle$ is a coherent mechanical state.

The total system is isolated, therefore, the evolution of its density operator is given by the equation

$$\dot{\rho}_{\text{tot}}(t) = -\frac{\text{i}}{\hbar}[H_{\text{tot}}, \rho_{\text{tot}}(t)], \tag{3.24}$$

which becomes in the interaction picture

$$\dot{\rho}^{\text{I}}_{\text{tot}}(t) = -\frac{\text{i}}{\hbar}[V^{\text{I}}(t), \rho^{\text{I}}_{\text{tot}}(t)]. \tag{3.25}$$

Any operator $A(t)$ in the Schrödinger picture becomes in the interaction picture

$$A^{\text{I}}(t) = \exp\left(\frac{\text{i}}{\hbar}H_0 t\right) A(t) \exp\left(-\frac{\text{i}}{\hbar}H_0 t\right), \tag{3.26}$$

with $H_0 = H_{\text{qm}} + H_{\mathcal{R}_\text{q}}$, in particular $V^{\text{I}}(t) = \sum_{l=\pm} R^{\text{I}}_l(t) \otimes \sigma^{\text{I}}_l(t)$ with

$$R^{\text{I}}_+(t) = \sum_k \hbar g_k a_k \exp(-\text{i}\omega_k t), \tag{3.27}$$

$$\sigma^{\text{I}}_+(t) = \exp\left(\frac{\text{i}}{\hbar}H_{\text{qm}} t\right) \sigma_+ \exp\left(-\frac{\text{i}}{\hbar}H_{\text{qm}} t\right). \tag{3.28}$$

Because of the presence of V_{qm} (Eq. (3.4)) in H_0, $\sigma^{\text{I}}_\pm$ also acts on the MO. Integrating Eq. (3.25) over one time step yields

$$\rho^{\text{I}}_{\text{tot}}(t+\Delta t) = \rho^{\text{I}}_{\text{tot}}(t) - \frac{\text{i}}{\hbar}\int_t^{t+\Delta t} \text{d}t'[V^{\text{I}}(t'), \rho^{\text{I}}_{\text{tot}}(t')]. \tag{3.29}$$

Replacing $\rho^{\text{I}}_{\text{tot}}(t')$ in the same way inside the integral gives

$$\begin{aligned}\Delta\rho^{\text{I}}_{\text{tot}}(t) &= \rho^{\text{I}}_{\text{tot}}(t+\Delta t) - \rho^{\text{I}}_{\text{tot}}(t) \\ &= -\frac{\text{i}}{\hbar}\int_t^{t+\Delta t} \text{d}t'[V^{\text{I}}(t'), \rho^{\text{I}}_{\text{tot}}(t)] \\ &\quad - \frac{1}{\hbar^2}\int_t^{t+\Delta t} \text{d}t' \int_t^{t'} \text{d}t''[V^{\text{I}}(t'), [V^{\text{I}}(t''), \rho^{\text{I}}_{\text{tot}}(t'')]].\end{aligned} \tag{3.30}$$

The total density operator at time t'' writes

$$\rho^{\text{I}}_{\text{tot}}(t'') = \rho^{\text{I}}_{\text{qm}}(t'') \otimes \rho^{\text{I}}_{\mathcal{R}_\text{q}}(t'') + \rho_{\text{corr}}(t'') \tag{3.31}$$

To derive the master equation for the optomechanical system, we trace over the reservoir's Hilbert and apply the *Born-Markov* approximations:

- Born approximation: The coupling between the qubit and the reservoir is weak therefore the state of $\mathcal{R}_\text{q}$ is only negligibly modified by the interaction and $\tau_\text{c} \ll \Delta t$ so the correlations vanish quickly and have a negligible impact on the evolution of

the optomechanical system. $\rho^{\mathrm{I}}_{\mathrm{tot}}(t')$ can thus be replaced by $\rho^{\mathrm{I}}_{\mathrm{qm}}(t') \otimes \rho_{\mathcal{R}_{\mathrm{q}}}$ where the state of the reservoir is time-independent in the interaction picture.
- Markov approximation: The time step is a lot shorter than the characteristic evolution time of the optomechanical system (Eq. (3.21)), so we can neglect its evolution between t and t'.

Moreover, the reservoir is in a thermal state, i.e. diagonal in the Fock state basis so $\mathrm{Tr}_{\mathcal{R}_{\mathrm{q}}}(R^{\mathrm{I}}_{\pm}(t)\rho_{\mathcal{R}_{\mathrm{q}}}) = 0$. The precursor of the master equation therefore reads

$$\begin{aligned}\Delta\rho^{\mathrm{I}}_{\mathrm{qm}}(t) &= \mathrm{Tr}_{\mathcal{R}_{\mathrm{q}}}(\Delta\rho^{\mathrm{I}}_{\mathrm{tot}}(t)) \\ &= -\frac{1}{\hbar^2}\int_t^{t+\Delta t} \mathrm{d}t' \int_t^{t'} \mathrm{d}t''\mathrm{Tr}_{\mathcal{R}_{\mathrm{q}}}\left(\left[V^{\mathrm{I}}(t'), \left[V^{\mathrm{I}}(t''), \rho^{\mathrm{I}}_{\mathrm{qm}}(t) \otimes \rho_{\mathcal{R}_{\mathrm{q}}}\right]\right]\right),\end{aligned} \tag{3.32}$$

where we have used that the first order term vanishes.

Then, expanding the commutators, the trace over the bath's degrees of freedom can be computed. It yields terms of the form

$$g_{ll'}(u, v) := \mathrm{Tr}_{\mathcal{R}_{\mathrm{q}}}(\rho_{\mathcal{R}_{\mathrm{q}}} R^{\mathrm{I}}_l(u)^\dagger R^{\mathrm{I}}_{l'}(v)), \tag{3.33}$$

where u and v are two times, and $l, l' \in \{+, -\}$. If $l \neq l'$ the trace vanishes, otherwise, we get the two correlation functions:

$$g_{--}(u, v) = \hbar^2 \sum_k g_k^2(\bar{n}_{\omega_k} + 1)\mathrm{e}^{-\mathrm{i}\omega_k(u-v)}, \tag{3.34}$$

$$g_{++}(u, v) = \hbar^2 \sum_k g_k^2\bar{n}_{\omega_k}\mathrm{e}^{\mathrm{i}\omega_k(u-v)}, \tag{3.35}$$

where $\bar{n}_{\omega_k}$ is the average number of photons of frequency ω_k in the bath. The integral $\int_t^{t'} \mathrm{d}t''$ can then be changed into an integral over $\tau = t' - t''$: $\int_0^{t'-t} \mathrm{d}\tau$. As $g_{ll}(u, v) = g_{ll}(u - v)$ is non zero only for $|u - v| \lesssim \tau_c \ll \Delta t$, the upper bound can be set to infinity [22]:

$$\begin{aligned}\Delta\rho^{\mathrm{I}}_{\mathrm{qm}}(t) = -\frac{1}{\hbar^2}\int_t^{t+\Delta t} \mathrm{d}t' \int_0^{\infty} \mathrm{d}\tau \sum_{l=\pm} g_{ll}(\tau)\big(&\sigma^{\mathrm{I}}_l(t')^\dagger\sigma^{\mathrm{I}}_l(t' - \tau)\rho^{\mathrm{I}}_{\mathrm{qm}}(t) \\ &- \sigma^{\mathrm{I}}_l(t' - \tau)\rho^{\mathrm{I}}_{\mathrm{qm}}(t)\sigma^{\mathrm{I}}_l(t')^\dagger\big) + \mathrm{h.c.}\end{aligned} \tag{3.36}$$

In addition, $\tau_{\mathrm{c}} \ll g_{\mathrm{m}}^{-1}$, so at the first order in $g_{\mathrm{m}}\tau$,

$$\sigma^{\mathrm{I}}_+(t' - \tau) = \sigma^{\mathrm{I}}_+(t')\mathrm{e}^{-\mathrm{i}\tilde{\omega}\tau}, \tag{3.37a}$$

$$\sigma^{\mathrm{I}}_-(t' - \tau) = \mathrm{e}^{\mathrm{i}\tilde{\omega}\tau}\sigma^{\mathrm{I}}_-(t'), \tag{3.37b}$$

where the operator $\tilde{\omega}$ is defined by

$$\tilde{\omega} := \omega_0 \mathbf{1}_{\mathrm{m}} + g_{\mathrm{m}} \left(b\mathrm{e}^{-\mathrm{i}\Omega t} + b^{\dagger}\mathrm{e}^{\mathrm{i}\Omega t} \right). \tag{3.38}$$

As $\Delta t \ll \gamma^{-1}, g_{\mathrm{m}}^{-1}, \Omega^{-1}$, integrating over t' approximately gives

$$\begin{aligned}
\dot{\rho}^{\mathrm{I}}_{\mathrm{qm}}(t) &= \frac{\Delta \rho^{\mathrm{I}}_{\mathrm{qm}}}{\Delta t}(t) \\
&= -\frac{1}{\hbar^2} \int_0^{\infty} \mathrm{d}\tau \Big[g_{++}(\tau) \big(\sigma^{\mathrm{I}}_{-}(t) \sigma^{\mathrm{I}}_{+}(t) \mathrm{e}^{-\mathrm{i}\tilde{\omega}\tau} \rho^{\mathrm{I}}_{\mathrm{qm}}(t) - \sigma^{\mathrm{I}}_{+}(t) \mathrm{e}^{-\mathrm{i}\tilde{\omega}\tau} \rho^{\mathrm{I}}_{\mathrm{qm}}(t) \sigma^{\mathrm{I}}_{-}(t) \big) \\
&\qquad + g_{--}(\tau) \big(\sigma^{\mathrm{I}}_{+}(t) \mathrm{e}^{\mathrm{i}\tilde{\omega}\tau} \sigma^{\mathrm{I}}_{-}(t) \rho^{\mathrm{I}}_{\mathrm{qm}}(t) - \mathrm{e}^{\mathrm{i}\tilde{\omega}\tau} \sigma^{\mathrm{I}}_{-}(t) \rho^{\mathrm{I}}_{\mathrm{qm}}(t) \sigma^{\mathrm{I}}_{+}(t) \big) \Big] \\
&\quad + \mathrm{h.c.}
\end{aligned} \tag{3.39}$$

The coupling to the bath solely induces transitions between the qubit's bare energy states, such that the hybrid system naturally evolves into a classically correlated state of the form $\rho_{\mathrm{qm}}(t) = P_e(t) \, |e\rangle\langle e| \otimes |\beta_e(t)\rangle\langle \beta_e(t)| + P_g(t) \, |g\rangle\langle g| \otimes \big|\beta_g(t)\big\rangle\big\langle\beta_g(t)\big|$. The state mechanical $\beta_\epsilon(t)$, $\epsilon \in \{e, g\}$, given by Eq. (3.12), can be rewritten

$$\beta_\epsilon = \beta_0 e^{-\mathrm{i}\Omega t} + \delta\beta_\epsilon(t), \tag{3.40}$$

where the mechanical fluctuations have no influence on the qubit frequency as long as $|\delta\beta_\epsilon(t)| \ll |\beta_0|$, i.e. $t \ll |\beta_0| g_{\mathrm{m}}^{-1}$. Therefore, as long as $t \ll |\beta_0| g_{\mathrm{m}}^{-1}$, $\omega(\beta_e(t)) \simeq \omega(\beta_g(t)) \simeq \omega(\beta_0(t))$, where

$$\omega(\beta) = \omega_0 + g_{\mathrm{m}}(\beta + \beta^*) \tag{3.41}$$

is the effective qubit's frequency when the MO is in state $|\beta\rangle$ and

$$\beta_0(t) = \beta_0 e^{-i\Omega t} \tag{3.42}$$

corresponds to the free evolution of the MO. Denoting

$$\big|E^{\mathrm{I}}(t)\big\rangle = \exp\left(\frac{\mathrm{i}}{\hbar} H_{\mathrm{qm}} t \right) |e\rangle \otimes |\beta_e(t)\rangle \,, \tag{3.43}$$

$$\big|G^{\mathrm{I}}(t)\big\rangle = \exp\left(\frac{\mathrm{i}}{\hbar} H_{\mathrm{qm}} t \right) |g\rangle \otimes \big|\beta_g(t)\big\rangle, \tag{3.44}$$

then, $\exp(\mathrm{i}\tilde{\omega}\tau)$ verifies

$$\left\langle G^{\mathrm{I}}(t) \left| \mathrm{e}^{\mathrm{i}\tilde{\omega}\tau} \sigma_{-}^{\mathrm{I}}(t) \right| E^{\mathrm{I}}(t) \right\rangle \simeq \mathrm{e}^{\mathrm{i}\omega(\beta_0(t))\tau} \tag{3.45}$$

$$\begin{aligned} \left\langle G^{\mathrm{I}}(t) \left| \mathrm{e}^{\mathrm{i}\tilde{\omega}\tau} \sigma_{-}^{\mathrm{I}}(t) \right| G^{\mathrm{I}}(t) \right\rangle &= \left\langle E^{\mathrm{I}}(t) \left| \mathrm{e}^{\mathrm{i}\tilde{\omega}\tau} \sigma_{-}^{\mathrm{I}}(t) \right| E^{\mathrm{I}}(t) \right\rangle \\ &= \left\langle E^{\mathrm{I}}(t) \left| \mathrm{e}^{\mathrm{i}\tilde{\omega}\tau} \sigma_{-}^{\mathrm{I}}(t) \right| G^{\mathrm{I}}(t) \right\rangle \\ &= 0. \end{aligned} \tag{3.46}$$

$\dot{\rho}_{\mathrm{qm}}^{\mathrm{I}}(t)$ can then be decomposed over the states $\left|E^{\mathrm{I}}(t)\right\rangle$, $\left|G^{\mathrm{I}}(t)\right\rangle$ and the integral over τ make the system interacts only with bath photons of frequency $\omega(\beta_0(t))$.

The master equation describing the relaxation of the hybrid system in the bath, in the Schrödinger picture, can finally be written as

$$\begin{aligned} \dot{\rho}_{\mathrm{qm}}(t) = &-\frac{\mathrm{i}}{\hbar}[H_{\mathrm{qm}}, \rho_{\mathrm{qm}}(t)] + \gamma_{\omega(\beta_0(t))}\bar{n}_{\omega(\beta_0(t))} D[\sigma_+ \otimes \mathbf{1}_{\mathrm{m}}]\rho_{\mathrm{qm}}(t) \\ &+ \gamma_{\omega(\beta_0(t))}\left(\bar{n}_{\omega(\beta_0(t))} + 1\right) D[\sigma_- \otimes \mathbf{1}_{\mathrm{m}}]\rho_{\mathrm{qm}}(t). \end{aligned} \tag{3.47}$$

This equation is valid in the semi-classical regime $t \ll |\beta_0| g_{\mathrm{m}}^{-1}$. The spontaneous emission rate $\gamma_{\omega(\beta_0(t))}$ of the qubit, given by Eq. (3.16), and the average number of photons $\bar{n}_{\omega(\beta_0(t))}$, given by Eq. (3.14), depend on the effective transition frequency of the qubit $\omega(\beta_0(t))$. Furthermore, the Hamiltonian (3.4) is a linear approximation [16] of the optomechanical coupling, so to remain in the domain of validity of this approximation we will consider mechanical amplitudes β such that

$$|\omega(\beta) - \omega_0| \ll \omega_0. \tag{3.48}$$

and assume in the rest of this dissertation that

$$\gamma_{\omega(\beta)} = \gamma_{\omega_0} = \gamma. \tag{3.49}$$

3.2 Thermodynamics

This section presents the average thermodynamics of the qubit. First, we study adiabatic transformations, i.e. an isolated hybrid optomechanical system, because there is no heat exchanged which makes it easier to identify work. Secondly, we consider isothermal transformations, using the master equation derived above.

3.2.1 Adiabatic Transformations

Here we consider an adiabatic transformation in the thermodynamic sense of the term. We will therefore use the definitions and results from Sect. 3.1.1.1. Using Eq. (2.21), we define the energy of the hybrid optomechanical system

$$\varepsilon_{\mathrm{qm}}(t) = \mathrm{Tr}_{\mathrm{qm}}(\rho_{\mathrm{qm}}(t) H_{\mathrm{qm}}). \tag{3.50}$$

The evolution of the isolated hybrid optomechanical system is given by

$$\dot{\rho}_{\mathrm{qm}}(t) = -\frac{\mathrm{i}}{\hbar}[H_{\mathrm{qm}}, \rho_{\mathrm{qm}}(t)]. \tag{3.51}$$

Taking the partial trace over the mechanics, we obtain the evolution of the qubit

$$\dot{\rho}_{\mathrm{q}}(t) = -\frac{\mathrm{i}}{\hbar}[H_{\mathrm{q}}^{\mathrm{eff}}(t), \rho_{\mathrm{q}}(t)], \tag{3.52}$$

where we have introduced the effective Hamiltonian

$$H_{\mathrm{q}}^{\mathrm{eff}}(t) := \mathrm{Tr}_{\mathrm{m}}(\rho_{\mathrm{m}}(t)(H_{\mathrm{q}} + V_{\mathrm{qm}})) = \hbar \left(\omega_0 + g_{\mathrm{m}} \frac{x(t)}{x_{\mathrm{zpf}}} \right) |e\rangle\langle e| \,, \tag{3.53}$$

with $x(t) = \mathrm{Tr}_{\mathrm{m}}(\rho_{\mathrm{m}}(t)\hat{x})$. The motion of the MO thus results in an effective time-dependent Hamiltonian acting on the qubit, which is reminiscent of the action of the battery described in Chap. 2. Therefore we use Eq. (2.24) to define the work rate received by the qubit

$$\dot{W}(t) := \mathrm{Tr}_{\mathrm{q}}(\rho_{\mathrm{q}}(t)\dot{H}_{\mathrm{q}}^{\mathrm{eff}}(t)) = \hbar g_{\mathrm{m}} \frac{\dot{x}(t)}{x_{\mathrm{zpf}}} P_e. \tag{3.54}$$

We have defined the population of the excited state

$$P_e := \mathrm{Tr}_{\mathrm{q}}(\rho_{\mathrm{q}}(t)\, |e\rangle\langle e|), \tag{3.55}$$

which is time-independent in the adiabatic case because $|e\rangle\langle e|$ commutes with $H_{\mathrm{q}}^{\mathrm{eff}}(t)$. The total work received by the qubit is therefore

$$W = \hbar g_{\mathrm{m}} \frac{\Delta x}{x_{\mathrm{zpf}}} P_e = \hbar \Delta\omega P_e, \tag{3.56}$$

with $\Delta x = x(t_{\mathrm{f}}) - x(t_0)$ and $\Delta\omega$ the variation of the qubit's transition frequency. The physical interpretation of this work is the following: When the qubit is in the ground state, its energy is always zero, so it does not cost any work to change the frequency of the qubit's transition. However, when the qubit is excited, increasing the frequency of the qubit's transition requires to increase the qubit's energy so an equivalent amount of work has to be provided.

In addition, using Eq. (3.51), we get $\dot{\varepsilon}_{\rm qm}(t) = 0$, as expected due to energy conservation. Therefore, splitting the Hamiltonian $H_{\rm qm}$, we obtain

$$\begin{aligned} 0 &= {\rm Tr}_{\rm q}(\dot{\rho}_{\rm q}(t) H_{\rm q}) + {\rm Tr}_{\rm m}(\dot{\rho}_{\rm m}(t) H_{\rm m}) + {\rm Tr}_{\rm qm}(\dot{\rho}_{\rm qm}(t) V_{\rm qm}) \\ &= 0 + {\rm Tr}_{\rm m}(\dot{\rho}_{\rm m}(t) H_{\rm m}) + \hbar g_{\rm m} \frac{\dot{x}(t)}{x_{\rm zpf}} P_e \\ &= \dot{\varepsilon}_{\rm m}(t) + \dot{W}(t), \end{aligned} \tag{3.57}$$

where we have used Eq. (3.54) and defined the mechanical energy as

$$\varepsilon_{\rm m}(t) := {\rm Tr}_{\rm m}(\rho_{\rm m}(t) H_m). \tag{3.58}$$

Finally, the integration of Eq. (3.57) between t_0 and $t_{\rm f}$ yields

$$W = \varepsilon_{\rm m}(t_0) - \varepsilon_{\rm m}(t_{\rm f}) = -\Delta\varepsilon_{\rm m}. \tag{3.59}$$

Therefore, the MO provides work to the qubit during the transformation and clearly plays the role of the battery. Moreover, since the MO is not a large classical system but a nanomechanical resonator, its energy change $\Delta\varepsilon_{\rm m}$ is non negligible and therefore potentially measurable, as explained below. This gives us a direct way to access work by measuring the energy of the MO at the start and end of the transformation.

To get a clearer picture of the MO as both a battery and a work meter, we will now focus on the simple case were the hybrid optomechanical system is initially prepared in the state $\rho_{\rm qm}(0) = |\epsilon, \beta_0\rangle\langle\epsilon, \beta_0|$, with $|\epsilon, \beta_0\rangle = |\epsilon\rangle \otimes |\beta_0\rangle$ and $\epsilon \in \{e, g\}$. So, at time t, the state of the optomechanical system reads

$$\rho_{\rm qm}(t) = |\epsilon, \beta_\epsilon(t)\rangle\langle\epsilon, \beta_\epsilon(t)| \,, \tag{3.60}$$

namely the qubit's state does not change while the MO evolves according to Eq. (3.12). The evolution of the MO is represented in phase space defined by the mean quadratures $X = \langle b + b^\dagger \rangle /2$ and $P = -i \langle b - b^\dagger \rangle /2$ in Fig. 3.2b. The qubit's transition frequency is $\omega(\beta_\epsilon(t))$, so as long as $|\beta_0| \gg g_{\rm m}/\Omega$, both frequencies are almost identical: $\omega(\beta_e(t)) \simeq \omega(\beta_g(t))$. Therefore, the transformation undergone by the qubit, represented in Fig. 3.2a, does not depend on its state, as with an external drive. On the other hand, the mechanical energy is given by $\varepsilon_{\rm m}(t) = \hbar\Omega|\beta_\epsilon(t)|^2$. After a quarter of a mechanical period, $|\beta_e(t_{\rm f}) - \beta_g(t_{\rm f})| = \sqrt{2} g_{\rm m}/\Omega$ with $t_{\rm f} = \pi/2\Omega$. In the ultra-strong coupling regime, we have $|\beta_e(t_{\rm f}) - \beta_g(t_{\rm f})| > 1$, i.e. this difference is larger than the zero-point fluctuations, therefore it is measurable. As a consequence, the work received by the qubit can be directly obtained by measuring the mechanical energy variation. In practice, the mechanical energy can be computed from a time resolved measurement of the position $x(t)$ of the MO [23, 24].

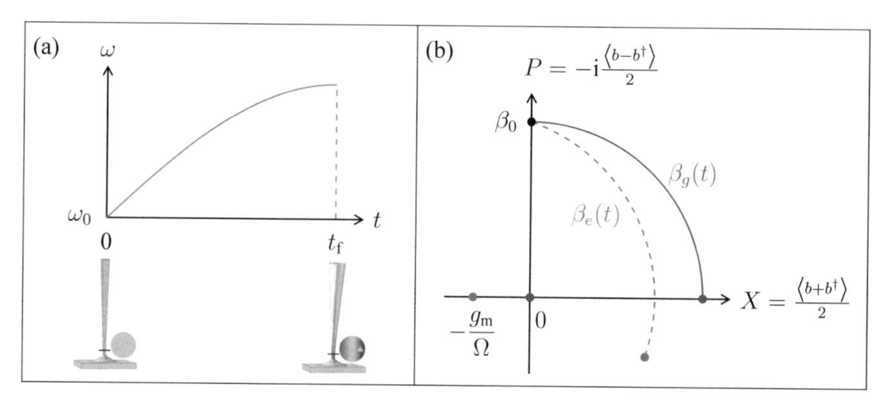

Fig. 3.2 **a** Modulation of the frequency of the qubit by the MO. **b** Evolution of the MO for each energy eigenstate of the qubit (dashed red line for $|e\rangle$ and solid blue line for $|g\rangle$)

3.2.2 Isothermal Transformations

We now add the qubit's bath $\mathcal{R}_\mathrm{q}$. The transformation is therefore isothermal. In the following, we restrict ourselves to the regime in which the master equation (3.47) is valid, i.e. on transformations whose duration t_f fulfill $t_\mathrm{f} \ll |\beta_0| g_\mathrm{m}^{-1}$. Product states of the form $\rho_\mathrm{qm}(t) = \rho_\mathrm{q}(t) \otimes \rho_\mathrm{m}(t)$ are natural solutions of Eq. (3.47), giving rise to two reduced coupled equations respectively governing the dynamics of the qubit and the mechanics:

$$\dot{\rho}_\mathrm{q}(t) = -\frac{\mathrm{i}}{\hbar}[H_\mathrm{q}^\mathrm{eff}(t), \rho_\mathrm{q}(t)] + \mathcal{L}_\mathrm{q}^t[\rho_\mathrm{q}(t)], \tag{3.61}$$

$$\dot{\rho}_\mathrm{m}(t) = -\frac{\mathrm{i}}{\hbar}[H_\mathrm{m}^\mathrm{eff}(t), \rho_\mathrm{m}(t)], \tag{3.62}$$

where the Lindbladian of the qubit reads

$$\mathcal{L}_\mathrm{q}^t[\rho_\mathrm{q}(t)] = \gamma \bar{n}_{\omega(\beta_0(t))} D[\sigma_+]\rho_\mathrm{q}(t) + \gamma\left(\bar{n}_{\omega(\beta_0(t))} + 1\right) D[\sigma_-]\rho_\mathrm{q}(t). \tag{3.63}$$

These equations are identical to the ones that were derived in Ref. [15] in the regime $\omega_0 \gg \gamma \gg g_\mathrm{m} \geq \Omega$. We have introduced the effective time-dependent Hamiltonians

$$H_\mathrm{q}^\mathrm{eff}(t) = \mathrm{Tr}_\mathrm{m}(\rho_\mathrm{m}(t)(H_\mathrm{q} + V_\mathrm{qm})) = \hbar\omega(\beta_0(t))\,|e\rangle\langle e|\,, \tag{3.64}$$

$$H_\mathrm{m}^\mathrm{eff}(t) = \mathrm{Tr}_\mathrm{q}(\rho_\mathrm{q}(t)(H_\mathrm{m} + V_\mathrm{qm})). \tag{3.65}$$

These semi-classical equations physically mean that the qubit exerts a force on the MO, resulting in the effective Hamiltonian $H_\mathrm{m}^\mathrm{eff}(t)$ while the MO modulates the frequency $\omega(\beta_0(t))$ of the qubit (Eq. (3.41)), which makes the coupling parameters of the qubit to the bath time-dependent. This is similar to the situation described in the previous section, except that there are now heat exchanges between the qubit and its bath.

Like in Sect. 2.2.1, we define the internal energy of the qubit and the heat rate as

$$\varepsilon_{\mathrm{q}}(t) := \mathrm{Tr}(\rho_{\mathrm{q}}(t) H_{\mathrm{q}}^{\mathrm{eff}}(t)), \tag{3.66}$$

$$\dot{Q}(t) := \mathrm{Tr}(\dot{\rho}_{\mathrm{q}}(t) H_{\mathrm{q}}^{\mathrm{eff}}(t)). \tag{3.67}$$

The work rate received by the qubit is still given by Eq. (3.54), therefore the work and heat received by the qubit during the transformation read

$$W = \int_0^{t_{\mathrm{f}}} \mathrm{d}t\, \hbar g_{\mathrm{m}} \frac{\dot{x}(t)}{x_{\mathrm{zpf}}} P_e(t), \tag{3.68}$$

$$Q = \int_0^{t_{\mathrm{f}}} \mathrm{d}t \mathrm{Tr}(\mathcal{L}_{\mathrm{q}}^t[\rho_{\mathrm{q}}(t)] H_{\mathrm{q}}^{\mathrm{eff}}(t)). \tag{3.69}$$

They fulfill the first law

$$\Delta\varepsilon_{\mathrm{q}} = W + Q. \tag{3.70}$$

We also define the internal energy of the whole hybrid optomechanical system as

$$\varepsilon_{\mathrm{qm}}(t) := \mathrm{Tr}(\rho_{\mathrm{q}}(t) \otimes \rho_{\mathrm{m}}(t) H_{\mathrm{qm}}), \tag{3.71}$$

which naturally splits into $\varepsilon_{\mathrm{qm}}(t) = \varepsilon_{\mathrm{q}}(t) + \varepsilon_{\mathrm{m}}(t)$. The total energy variation reads

$$\Delta\varepsilon_{\mathrm{qm}} = \Delta\varepsilon_{\mathrm{q}} + \Delta\varepsilon_{\mathrm{m}} = Q, \tag{3.72}$$

where we have used Eq. (3.69). From this result and the first law for the qubit (3.70), we obtain

$$W = -\Delta\varepsilon_{\mathrm{m}}, \tag{3.73}$$

like in the adiabatic case and in Ref. [15]. This confirms that the MO provides all the work received by the qubit and therefore behaves as a proper battery. The result (3.73) is important because it means that the average work received by the system, which depends on the full evolution, can be directly read out by measuring the energy of the MO at the beginning and at the end of the transformation, there is no need to monitor either the system or the bath.

3.3 Summary

We presented the hybrid optomechanical systems that will be investigated further in Chaps. 4 and 5. They consist of a qubit coupled to a nanomechanical oscillator (MO). We focused on the ultra-strong coupling regime which is within reach of state-of-the-art experimental implementations. We first presented the Hamiltonian description of the system and of the qubit's bath which is a collection of harmonic oscillators in thermal equilibrium. We derived the master equation for the optomechanical system from this microscopic description.

Secondly, we studied the thermodynamics of the qubit. We considered first the adiabatic case where there are no heat exchanges, then the isothermal case, with the qubit's bath, in the semi-classical regime were the qubit and MO states can be factorized. We showed that in both cases, the MO behaves like a battery, providing work to the qubit. Besides, the finite size of the MO allows a direct measurement of the work received by the qubit by measuring the mechanical energy variation. This kind of measurement is not possible for a classical battery that is not impacted by its coupling to the system due to its large size. Furthermore, the Hamiltonian of the total hybrid optomechanical system is time-independent, therefore this device can be seen as an autonomous thermal machine. All these results evidence that hybrid optomechanical systems are promising test-beds to experimentally explore the thermodynamics of a single qubit.

References

1. Braginski V, Manukin A (1967) Ponderomotive effects of electromagnetic radiation. Sov Phys JETP 25(4):653–655
2. Dorsel A, McCullen JD, Meystre P, Vignes E, Walther H (1983) Optical bistability and mirror confinement induced by radiation pressure. Phys Rev Lett 51(17):1550–1553. https://doi.org/10.1103/PhysRevLett.51.1550
3. Schreppler S, Spethmann N, Brahms N, Botter T, Barrios M, Stamper- Kurn DM (2014) Optically measuring force near the standard quantum limit. Science 344(6191):1486–1489. https://doi.org/10.1126/science.1249850
4. Verlot P, Tavernarakis A, Briant T, Cohadon P-F, Heidmann A (2010) Backaction amplification and quantum limits in optomechanical measurements. Phys Rev Lett 104(13):133 602. https://doi.org/10.1103/PhysRevLett.104.133602
5. Metzger CH, Karrai K (2004) Cavity cooling of a microlever. Nature 432:1002–1005. https://doi.org/10.1038/nature03118
6. Gigan S, Bohm HR, Paternostro M, Blaser F, Langer G, Hertzberg JB, Schwab KC, Bauerle D, Aspelmeyer M, Zeilinger A (2006) Self-cooling of a micromirror by radiation pressure. Nature 444(7115):67–70. https://doi.org/10.1038/nature05273
7. Arcizet O, Cohadon P-F, Briant T, Pinard M, Heidmann A (2006) Radiationpressure cooling and optomechanical instability of a micromirror. Nature 444(7115):71–74. https://doi.org/10.1038/nature05244
8. Vitali D, Gigan S, Ferreira A, Böhm HR, Tombesi P, Guerreiro A, Vedral V, Zeilinger A, Aspelmeyer M (2007) Optomechanical entanglement between a movable mirror and a cavity field. Phys Rev Lett 98(3):030 405. https://doi.org/10.1103/PhysRevLett.98.030405
9. Schmidt M, Ludwig M, Marquardt F (2012) Optomechanical circuits for nanomechanical continuous variable quantum state processing. New J Phys 14(12):125 005. https://doi.org/10.1088/1367-2630/14/12/125005
10. Grudinin IS, Lee H, Painter O, Vahala KJ (2010) Phonon laser action in a tunable two-level system. Phys Rev Lett 104(8):083 901. https://doi.org/10.1103/PhysRevLett.104.083901
11. Khurgin JB, Pruessner MW, Stievater TH, Rabinovich WS (2012) Laser-rate- equation description of optomechanical oscillators. Phys Rev Lett 108(22):223 904. https://doi.org/10.1103/PhysRevLett.108.223904

12. Jing H, Özdemir SK, Lü X-Y, Zhang J, Yang L, Nori F (2014) PT-symmetric phonon laser. Phys Rev Lett 113(5):053 604. https://doi.org/10.1103/PhysRevLett.113.053604
13. Jiang Y, Maayani S, Carmon T, Nori F, Jing H (2018) Nonreciprocal phonon laser. Phys. Rev. Applied 10(6):064 037. https://doi.org/10.1103/PhysRevApplied.10.064037
14. Zhang Y-L, Zou C-L, Yang C-S, Jing H, Dong C-H, Guo G-C, Zou X-B (2018) Phase-controlled phonon laser. New J Phys 20(9):093 005. https://doi.org/10.1088/1367-2630/aadc9f
15. Elouard C, Richard M, Auffèves A (2015) Reversible work extraction in a hybrid optomechanical system. New J Phys 17(5):055 018. https://doi.org/10.1088/1367-2630/17/5/055018
16. Treutlein P, Genes C, Hammerer K, Poggio M, Rabl P (2014) Hybrid mechanical systems. In: Aspelmeyer M, Kippenberg TJ, Marquardt F (eds) Cavity optomechanics: nano- and micromechanical resonators interacting with light. Quantum science and technology. Springer, Berlin, pp 327–351
17. LaHaye MD, Suh J, Echternach PM, Schwab KC, Roukes ML (2009) Nanomechanical measurements of a superconducting qubit. Nature 459(7249):960–964. https://doi.org/10.1038/nature08093
18. Pirkkalainen J-M, Cho SU, Li J, Paraoanu GS, Hakonen PJ, Sillanpää MA (2013) Hybrid circuit cavity quantum electrodynamics with a micromechanical resonator. Nature 494(7436):211–215. https://doi.org/10.1038/nature11821
19. Arcizet O, Jacques V, Siria A, Poncharal P, Vincent P, Seidelin S (2011) A single nitrogen-vacancy defect coupled to a nanomechanical oscillator. Nat Phys 7(11):879. https://doi.org/10.1038/nphys2070
20. Yeo I, de Assis P-L, Gloppe A, Dupont-Ferrier E, Verlot P, Malik NS, Dupuy E, Claudon J, Gérard J-M, Auffèves A, Nogues G, Seidelin S, Poizat J-P, Arcizet O, Richard M (2014) Strain-mediated coupling in a quantum dot-mechanical oscillator hybrid system. Nat Nanotech 9:106–110. https://doi.org/10.1038/nnano.2013.274
21. Elouard C (2017) Thermodynamics of quantum open systems : applications in quantum optics and optomechanics, PhD thesis, Université Grenoble Alpes
22. Cohen-Tannoudji C, Dupont-Roc J, Grynberg G (2004) Atom-photon interactions: basic processes and applications. Physics textbook. Wiley, Weinheim-VCH
23. Sanii B, Ashby PD (2010) High sensitivity deflection detection of nanowires. Phys Rev Lett 104(14):147 203. https://doi.org/10.1103/PhysRevLett.104.147203
24. Mercier de Lépinay L, Pigeau B, Besga B, Vincent P, Poncharal P, Arcizet O (2017) A universal and ultrasensitive vectorial nanomechanical sensor for imaging 2D force fields. Nat Nanotechnol 12(2):156–162. https://doi.org/10.1038/nnano.2016.193

Chapter 4
Stochastic Thermodynamics of Hybrid Optomechanical Systems

The ability to define and measure entropy production in the quantum regime is key to optimizing quantum heat engines and minimizing the energetic cost of quantum information technologies [1–4]. Many fluctuations theorems, like Jarzynski equality (JE) [5, 6], have been generalized to quantum systems. However, measuring a quantum fluctuation theorem can be problematic in the genuinely quantum situation of a coherently driven quantum system, because of the fundamental and practical issues to define and measure quantum work mentioned in introduction [7–10].

In particular, JE has been experimentally verified only for quantum closed systems, that is systems that are driven but otherwise isolated, for instance trapped ions [11, 12], ensemble of cold atoms [13], and spins in Nuclear Magnetic Resonance (NMR) [14]. Therefore, new experimentally realistic strategies need to be developed to measure the fluctuations of entropy production for quantum open systems. Since work is usually provided by a classical operator, like in Chap. 2, most proposals are based on the measurement of heat fluctuations, obtained by monitoring the bath. This requires to engineer the bath and to develop high efficiency detection schemes [15–17] and no experimental demonstration has been conducted so far.

In this chapter, we propose an alternative, and experimentally feasible, strategy to measure the thermodynamic arrow of time for a quantum open system in Jarzynski's protocol. This strategy is based on the direct measurement of work fluctuations. In Chap. 3, we have seen that hybrid optomechanical systems are promising platforms for experimental quantum thermodynamics because the average work exchanges can be obtained by measuring the mechanical oscillator (MO). We now go one step further and show that work fluctuations equal the mechanical energy fluctuations, providing a direct way to access the stochastic entropy production. We first focus on the qubit and prove that its work fluctuations verify JE. Then, we consider the whole hybrid system which verifies a generalized integral fluctuation theorem (IFT) involving the information encoded in the battery. This work is published in [18].

J. Monsel, *Quantum Thermodynamics and Optomechanics*, Springer Theses,
https://doi.org/10.1007/978-3-030-54971-8_4

4.1 Quantum Trajectories

We consider the same situation as in the previous chapter: a hybrid optomechanical system whose qubit is also coupled to a thermal reservoir $\mathcal{R}_q$, as depicted in Fig. 4.1a, in the regime detailed in Sect. 3.1.2. To go the single realization level, we will unravel the master equation (3.47). In this chapter, the hybrid optomechanical system is initially in the state $\rho_{qm}(t_0) = \rho_q^\infty(\beta_0) \otimes |\beta_0\rangle\langle\beta_0|$, with

$$\rho_q(\beta_0) = p_{\beta_0}^\infty[e]\,|e\rangle\langle e| + p_{\beta_0}^\infty[g]\,|g\rangle\langle g|\,. \tag{4.1}$$

$\rho_q(\beta_0)$ is the thermal equilibrium state of the qubit at frequency $\omega(\beta_0)$, defined by Eq. (3.41), with

$$p_\beta^\infty[\epsilon] = \frac{1}{Z(\beta)}\exp(-\frac{\hbar\omega(\beta)\delta_{\epsilon,e}}{k_B T}), \tag{4.2}$$

and $Z(\beta)$ the partition function of the qubit, that reads

$$Z(\beta) = 1 + \exp(-\frac{\hbar\omega(\beta)}{k_B T}). \tag{4.3}$$

We study the evolution of the optomechanical system between times t_0 and t_f and, to stay in the regime of validity of the master equation, we assume that $g_m t_f \ll |\beta_0|$.

4.1.1 Direct Protocol

To obtain quantum trajectories, we apply a quantum jump unraveling to the master equation (3.47). The evolution of the hybrid optomechanical system between the times t_0 and $t_f = N\Delta t$ is therefore described by a stochastic trajectory $\vec{\Sigma} = \{|\Psi_\Sigma(t_n)\rangle\}_{n=0}^N$, where $|\Psi_\Sigma(t_n)\rangle$ is a vector in the optomechanical Hilbert space and $t_n = n\Delta t$ where Δt is the same time increment as in the previous chapter, fulfilling the criterion (3.21). The Kraus operators associated to this unraveling are

$$M_0(t_n) = \mathbf{1}_{qm} - \frac{i\Delta t}{\hbar}H_{eff}(t_n), \tag{4.4a}$$

$$M_+(t_n) = \sqrt{\gamma\Delta t\bar{n}_{\omega(\beta_0(t_n))}}\;\sigma_+ \otimes \mathbf{1}_m, \tag{4.4b}$$

$$M_-(t_n) = \sqrt{\gamma\Delta t(\bar{n}_{\omega(\beta_0(t_n))} + 1)}\;\sigma_- \otimes \mathbf{1}_m. \tag{4.4c}$$

$\mathbf{1}_{qm} = \mathbf{1}_q \otimes \mathbf{1}_m$ denotes the identity operator in the optomechanical Hilbert space. $\beta_0(t)$ is the free evolution of the MO, given by Eq. (3.42). M_- and M_+ are the jump operators. They experimentally correspond to the emission or absorption of a photon in the bath, associated with the transition of the qubit in the ground or excited state respectively while the state of the MO remains unchanged. Reciprocally, the

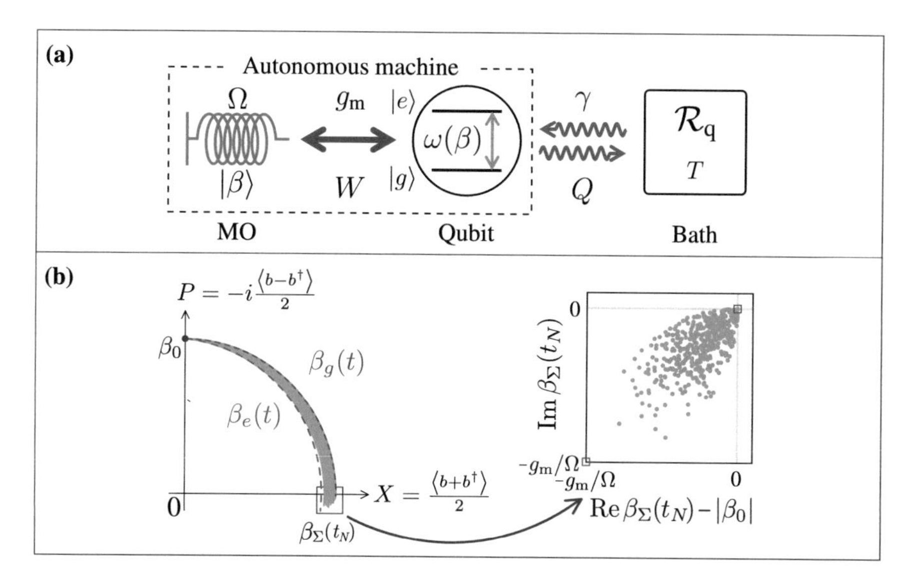

Fig. 4.1 **a** Situation under study: A qubit exchanges work W with a MO and heat Q with a thermal reservoir $\mathcal{R}_q$ at temperature T. The ensemble of the qubit and mechanics, i.e. the hybrid optomechanical system, constitutes an autonomous machine. **b** Stochastic trajectories of the MO $\beta[\vec{\epsilon}\,]$ in phase space. The MO is initially prepared in the coherent state $\big|\mathrm{i}|\beta_0|\big\rangle$ and the qubit state is drawn from thermal equilibrium. Inset: Distribution of final states $|\beta_\Sigma(t_N)\rangle$ within an area of typical width g_m/Ω. The blue and red squares indicates $\beta_g(t_N)$ and $\beta_e(t_N)$ respectively. Parameters: $T = 80$ K, $\hbar\omega_0 = 1.2k_BT$, $\Omega/2\pi = 100$ kHz, $\gamma/\Omega = 5$, $g_m/\Omega = 100$, $|\beta_0| = 1000$

no-jump operator M_0 corresponds to the absence of detection event in the bath, i.e. a continuous, non-Hermitian evolution governed by the effective Hamiltonian (Eq. (2.14))

$$H_{\mathrm{eff}}(t_n) = H_{\mathrm{qm}} - \mathrm{i}\hbar\frac{\gamma\Delta t}{2}\Big((\bar{n}_{\omega(\beta_0(t_n))} + 1)\,|e\rangle\langle e| + \bar{n}_{\omega(\beta_0(t_n))}\,|g\rangle\langle g|\Big) \otimes \mathbf{1}_{\mathrm{m}}. \quad (4.5)$$

The initial state is randomly chosen from the product state $\rho_q(\beta_0) \otimes |\beta_0\rangle\langle\beta_0|$. Therefore, the initial state of the optomechanical system is $|\Psi_0\rangle = |\epsilon_0, \beta_0\rangle$ where $\epsilon_0 \in \{e, g\}$. Then, the trajectory $\vec{\Sigma}$ is fully defined by the stochastic sequence of jump / no-jump $\{r_\Sigma(t_n)\}_{n=0}^{N-1}$, with $r_\Sigma(t_n) \in \{0, +, -\}$, and the final state reads [19]

$$|\Psi_\Sigma(t_N)\rangle = \frac{1}{\sqrt{P[\vec{\Sigma}|\Psi_0]}}\prod_{n=0}^{N-1} M_{r_\Sigma(t_n)}\,|\Psi_0\rangle \quad (4.6)$$

where

$$P[\vec{\Sigma}|\Psi_0] = \prod_{n=0}^{N-1} P[\Psi_\Sigma(t_{n+1})|\Psi_\Sigma(t_n)], \quad (4.7)$$

is the probability of the trajectory $\vec{\Sigma}$ knowing that the initial state is $|\Psi_\Sigma(t_0)\rangle = |\Psi_0\rangle$. We have introduced

$$P[\Psi_\Sigma(t_{n+1})|\Psi_\Sigma(t_n)] = \left\langle \Psi_\Sigma(t_n) \left| M^\dagger_{r_\Sigma(t_n)} M_{r_\Sigma(t_n)} \right| \Psi_\Sigma(t_n) \right\rangle \tag{4.8}$$

the probability of the transition from $|\Psi_\Sigma(t_n)\rangle$ to $|\Psi_\Sigma(t_{n+1})\rangle$ during the n-th time step (Eq. (2.12)). The probability of the trajectory $\vec{\Sigma}$ reads

$$P[\vec{\Sigma}] = p^\infty_{\beta_0}[\epsilon_0] \prod_{n=0}^{N-1} P[\Psi_\Sigma(t_{n+1})|\Psi_\Sigma(t_n)], \tag{4.9}$$

and the density operator of the optomechanical system, solution of (3.47), is recovered by averaging over the trajectories:

$$\rho_{\text{qm}}(t_N) = \sum_{\vec{\Sigma}} P[\vec{\Sigma}]\, |\Psi_\Sigma(t_N)\rangle\langle\Psi_\Sigma(t_N)| \,. \tag{4.10}$$

During the n-th time step, either a jump occurs or the system evolves under the action of the no-jump operator $M_0(t_n)$. In the former case, the mechanical state is preserved while the qubit is projected on either $|e\rangle$ or $|g\rangle$. In the latter case, the evolution of the system is governed by the effective Hamiltonian $H_{\text{eff}}(t_n)$, which can be split into the Hermitian part H_{qm} and the non-Hermitian part $H_{\text{nh}}(t_n) = H_{\text{eff}}(t_n) - H_{\text{qm}}$. $H_{\text{eff}}(t_n)$ is diagonal in the qubit's energy eigenbasis (See Eq. (4.5)), so if the qubit is in an energy eigenstate, its state will not change. $H_{\text{nh}}(t_n)$ is the identity over the mechanical Hilbert space, therefore the MO will evolve under the sole action of H_{qm} which preserves the coherent nature of the mechanical field (See Sect. 3.1.1.1). As the initial state of the system is $|\epsilon_0, \beta_0\rangle$, the total system always remains in a product state $|\Psi_\Sigma(t_n)\rangle = |\epsilon_\Sigma(t_n), \beta_\Sigma(t_n)\rangle$, with $\epsilon_\Sigma(t_n) \in \{e, g\}$ and $|\beta_\Sigma(t_n)\rangle$ is the coherent state of the MO given by

$$|\beta_\Sigma(t_n)\rangle = \prod_{k=0}^{n-1} \exp(-\mathrm{i}\Delta t H_{\text{m}}^{\epsilon_\Sigma(t_k)})\, |\beta_0\rangle \,, \tag{4.11}$$

where H_{m}^ϵ is defined by Eq. (3.7). As a consequence, we can split the optomechanical trajectory into the trajectory of the qubit $\vec{\epsilon} = \{|\epsilon_\Sigma(t_n)\rangle\}_{n=0}^N$ and the trajectory of the MO $\vec{\beta} = \{|\beta_\Sigma(t_n)\rangle\}_{n=0}^N$, so $\vec{\Sigma} = \{\vec{\epsilon}, \vec{\beta}[\vec{\epsilon}\,]\}$. This is a key result of this chapter because, thanks to the product state, there will be no ambiguity to define the energies of the qubit and the MO. We have used the notation $\vec{\beta}[\vec{\epsilon}\,]$ because the mechanical trajectory can be fully reconstructed from the trajectory of the qubit using Eq. (4.11). In the semi-classical regime considered here, the jump probabilities only depends on $\omega(\beta_0(t))$ so that the reduced evolution of the qubit is Markovian. Conversely, the state of the MO at any time t_n depends on the full trajectory of the qubit up to time t_n. Therefore, by measuring the state of the MO at time t_N, we obtain information on the whole trajectory of the qubit.

At the end of the protocol, the state of the optomechanical system, averaged over the trajectories, is given by Eq. (4.10). So, the reduced mechanical average state, defined by $\rho_\mathrm{m}(t_N) = \mathrm{Tr}_\mathrm{q}(\rho_\mathrm{qm}(t_N))$, is a discrete distribution of the final mechanical states $\{|\beta_\Sigma(t_N)\rangle\}_{\vec{\Sigma}}$. Denoting $p_\mathrm{m}[\beta_\mathrm{f}]$ the probability that the MO ends in the coherent state $|\beta_\mathrm{f}\rangle$, we can express the reduced mechanical average state as

$$\rho_\mathrm{m}(t_N) = \sum_{\beta_\mathrm{f}} p_\mathrm{m}[\beta_\mathrm{f}]\,|\beta_\mathrm{f}\rangle\langle\beta_\mathrm{f}|\,, \tag{4.12}$$

where

$$\sum_{\beta_\mathrm{f}} p_\mathrm{m}[\beta_\mathrm{f}] = 1. \tag{4.13}$$

Examples of numerically generated trajectories $\vec{\beta}[\vec{\epsilon}\,]$ are plotted in the phase space defined by the mean quadratures $X = \langle b + b^\dagger\rangle/2$ and $P = -i\langle b - b^\dagger\rangle/2$ in Fig. 4.1b. These trajectories were obtained by sampling the ensemble of possible direct trajectories using the jump and no-jump probabilities given by Eq. (4.8) [19]. The final mechanical amplitude can be split into

$$\beta_\Sigma(t_N) = \beta_0(t_N) + \delta\beta_\Sigma(t_N), \tag{4.14}$$

where $\delta\beta_\Sigma(t_N)$ corresponds to the mechanical fluctuations caused by the interaction with the qubit. As visible in the inset of the figure, the final mechanical states are contained inside an area of typical size g_m/Ω, so these fluctuations are of the order of g_m/Ω. The semi-classical regime ensures that $|\delta\beta_\Sigma(t_n)| \ll |\beta_0(t_n)|$, so that the qubit's transition frequency is not sensitive to the mechanical fluctuations and the reduced evolution of the qubit is Markovian. On the other hand, the ultra-strong coupling regime $g_\mathrm{m}/\Omega \gg 1$ makes these fluctuations large enough to be measurable. Both regimes are compatible, which is the key to our proposal to measure work fluctuations, as detailed below.

Interestingly, the stochastic trajectory model described above can be extended beyond the semi-classical regime $t \ll |\beta_0| g_\mathrm{m}^{-1}$. In this quantum jump picture, at any time t_n, the system is in a product state $|\epsilon_\Sigma(t_n), \beta_\Sigma(t_n)\rangle$, where $|\beta_\Sigma(t_n)\rangle$ is a coherent mechanical state. We can therefore derive a master equation describing the evolution of the system over the next time step of the form of Eq. (3.47), and unravel it with a set of trajectory dependent Kraus operators similar to Eq. (4.4), but where the effective frequency of the qubit is $\omega(\beta_\Sigma(t_n))$ instead of $\omega(\beta_0(t_n))$. In this more general situation, the mechanical fluctuations can no longer be neglected compared to $|\beta_0(t_n)|$, therefore Eq. (4.10) is no longer a solution of the semi-classical master equation. Furthermore the frequency modulation of the qubit is now trajectory dependent, which means that the jump probabilities are as well. Therefore the reduced trajectory of the qubit is no longer Markovian. As shown below the Markovianity of the reduced trajectory of the qubit is key to our proposal of measurement of JE, which is therefore limited to the semi-classical regime.

4.1.2 *Reversed Protocol*

Our goal is to study fluctuation theorems in the hybrid optomechanical system, so we will need to compute the entropy production, given by Eq. (2.50) which requires to define the time-reversed protocol. This protocol, defined between t_N and t_0, consists in time-reversing the unitaries while keeping the same stochastic map. This leads to the following time-dependent reversed Kraus operators [20–23]:

$$\tilde{M}_0(t_n) = \mathbf{1}_{\text{qm}} + \frac{\mathrm{i}\Delta t}{\hbar} H_{\text{eff}}^{\dagger}(t_n), \tag{4.15a}$$

$$\tilde{M}_-(t_n) = M_+(t_n), \tag{4.15b}$$

$$\tilde{M}_+(t_n) = M_-(t_n), \tag{4.15c}$$

The initial state of the reversed trajectory is obtained in the following way. The mechanical state $|\beta_\Sigma(t_N)\rangle$ is drawn from the final distribution of states $\{|\beta_{\text{f}}\rangle\}$ generated by the direct protocol with probability $p_{\text{m}}[\beta_{\text{f}}]$ while the state of the qubit $|\epsilon_\Sigma(t_N)\rangle$ is drawn from the thermal equilibrium corresponding to $\beta_\Sigma(t_N)$, with probability $p^{\infty}_{\beta_\Sigma(t_N)}[\epsilon_\Sigma(t_N)]$ (Eq. (4.2)). Therefore, the probability of the reversed trajectory $\overleftarrow{\Sigma}$ reads

$$\tilde{P}[\overleftarrow{\Sigma}] = p_{\text{m}}[\beta_\Sigma(t_N)] p^{\infty}_{\beta_\Sigma(t_N)}[\epsilon_\Sigma(t_N)] \prod_{n=N-1}^{0} \tilde{P}[\Psi_\Sigma(t_n)|\Psi_\Sigma(t_{n+1})]. \tag{4.16}$$

We have introduced the reversed transition probability at time t_n

$$\tilde{P}[\Psi_\Sigma(t_n)|\Psi_\Sigma(t_{n+1})] = \left\langle \Psi_\Sigma(t_{n+1}) \left| \tilde{M}^{\dagger}_{r_\Sigma(t_n)} \tilde{M}_{r_\Sigma(t_n)} \right| \Psi_\Sigma(t_{n+1}) \right\rangle. \tag{4.17}$$

4.2 Stochastic Thermodynamics

In the rest of this chapter, we focus on the following protocol: At time t_0, the optomechanical system is prepared in the state $\rho_{\text{qm}}(t_0) = \rho^{\infty}_{\text{q}}(\beta_0) \otimes |\beta_0\rangle\langle\beta_0|$. An energy measurement of the qubit is performed, preparing the initial state of the trajectory $|\Psi_\Sigma(t_0)\rangle = |\epsilon_0, \beta_0\rangle$. Then, the qubit is coupled to the bath and the evolution of the optomechanical system is studied over a mechanical quarter period, i.e. $t_N = \pi/2\Omega$. This situation can be studied from two different perspectives depending on the choice of thermodynamic system, defining two distinct transformations. If the studied thermodynamic system is the whole hybrid optomechanical system, then the transformation is a thermal relaxation toward equilibrium. The initial state $\rho_{\mathrm{qm}}(t_0)$ is not an equilibrium state of optomechanical system and H_{qm} is time-independent, so the energy exchanges are reduced to heat exchanges with the bath. On the other hand, if the considered thermodynamic system is the qubit, the transformation consist in driving the qubit out of equilibrium. The qubit evolves under the action of the time-dependent effective Hamiltonian $H_{\mathrm{q}}^{\mathrm{eff}}(t)$, the driving work being provided by the MO. In the semi-classical regime, $H_{\mathrm{q}}^{\mathrm{eff}}(t) = \hbar\omega(\beta_0(t))\,|e\rangle\langle e|$ does not depend on the trajectory and this situation corresponds to Jarzynski's protocol. In this section, we define the stochastic thermodynamic quantities involved in both transformations.

4.2.1 Energy Exchanges

Applying the definition (2.40), the stochastic energy of the optomechanical system reads

$$\varepsilon_{\mathrm{qm}}[\vec{\Sigma}, t_n] = \left\langle \epsilon_\Sigma(t_n), \beta_\Sigma(t_N) \left| H_{\mathrm{qm}} \right| \epsilon_\Sigma(t_n), \beta_\Sigma(t_N) \right\rangle, \tag{4.18}$$

which naturally splits into the sum of the qubit's energy $\varepsilon_{\mathrm{q}}[\vec{\Sigma}, t_n]$ and the mechanical energy $\varepsilon_{\mathrm{m}}[\vec{\Sigma}, t_n]$, given by

$$\varepsilon_{\mathrm{q}}[\vec{\Sigma}, t_n] = \hbar\omega(\beta_\Sigma(t_n))\delta_{\epsilon_\Sigma(t_n),e}\,|e\rangle\langle e|\,, \tag{4.19}$$

$$\varepsilon_{\mathrm{m}}[\vec{\Sigma}, t_n] = \hbar\Omega|\beta_\Sigma(t_n)|^2. \tag{4.20}$$

During the n-th time step, these internal energies can change in two different ways depending on the nature of the stochastic event, jump or no-jump. If a quantum jump occurs, the mechanical state remains unchanged. Therefore, the mechanical energy variation is zero while the energies of the qubit and total optomechanical system vary by the same amount: $\delta\varepsilon_{\mathrm{qm}}[\vec{\Sigma}, t_n] = \delta\varepsilon_{\mathrm{q}}[\vec{\Sigma}, t_n]$. Following the definitions from Sect. 2.3.1, this energy exchange corresponds to heat provided by the bath and denoted $\delta Q[\vec{\Sigma}, t_n]$. Conversely, the no-jump evolution preserves the state of the qubit while its energy eigenvalues evolve due to the optomechanical coupling. This energy change is thus identified with work, denoted $\delta W[\vec{\Sigma}, t_n]$, and verifies $\delta\varepsilon_{\mathrm{q}}[\vec{\Sigma}, t_n] = \delta W[\vec{\Sigma}, t_n]$. During this time step, the optomechanical system is energetically isolated, therefore $\delta\varepsilon_{\mathrm{qm}}[\vec{\Sigma}, t_n] = 0$ and the work increment compensates the mechanical energy variation: $\delta\varepsilon_{\mathrm{m}}[\vec{\Sigma}, t_n] = -\delta W[\vec{\Sigma}, t_n]$.

Finally, the total work $W[\vec{\Sigma}]$ and heat $Q[\vec{\Sigma}]$ received by the qubit along the trajectory are obtained by summing up the increments (Eqs. (2.44) and (2.45)). By construction of the work and heat increments, the first law for the qubit is verified,

$$\Delta\varepsilon_{\mathrm{q}}[\vec{\Sigma}] = W[\vec{\Sigma}] + Q[\vec{\Sigma}], \tag{4.21}$$

while the total optomechanical energy variation corresponds to the heat exchanged,

$$\Delta\varepsilon_{\mathrm{qm}}[\vec{\Sigma}] = Q[\vec{\Sigma}]. \tag{4.22}$$

As a consequence, the work received by the qubit is fully provided by the MO:

$$W[\vec{\Sigma}] = -\Delta\varepsilon_{\mathrm{m}}[\vec{\Sigma}], \tag{4.23}$$

which extends to the single trajectory level the results from Sect. 3.2.2. This last equation is the second key result of this chapter, evidencing that the MO behaves like a battery and an ideal work meter at the single trajectory level.

4.2.2 Entropy Production

We now derive the expression of the entropy production, defined by Eq. (2.50). Using Eqs. (4.9) and (4.16), we write

$$\begin{aligned} s_{\mathrm{irr}}[\vec{\Sigma}] &= \log\left(\frac{P[\vec{\Sigma}]}{\tilde{P}[\overleftarrow{\Sigma}]}\right) \\ &= \log\left(\frac{p^{\infty}_{\beta_0}[\epsilon_\Sigma(t_0)]}{p_{\mathrm{m}}[\beta_\Sigma(t_N)]p^{\infty}_{\beta_\Sigma(t_N)}[\epsilon_\Sigma(t_N)]}\frac{\prod_{n=0}^{N-1}P[\Psi_\Sigma(t_{n+1})|\Psi_\Sigma(t_n)]}{\prod_{n=0}^{N-1}\tilde{P}[\Psi_\Sigma(t_n)|\Psi_\Sigma(t_{n+1})]}\right). \end{aligned} \tag{4.24}$$

Then, from the expressions of the jump and no-jump operators, we obtain

$$\begin{aligned} \frac{P[\Psi_\Sigma(t_{n+1})|\Psi_\Sigma(t_n)]}{\tilde{P}[\Psi_\Sigma(t_n)|\Psi_\Sigma(t_{n+1})]} &= \frac{\left\langle\Psi_\Sigma(t_n)\middle|M^\dagger_{r_\Sigma(t_n)}M_{r_\Sigma(t_n)}\middle|\Psi_\Sigma(t_n)\right\rangle}{\left\langle\Psi_\Sigma(t_{n+1})\middle|\tilde{M}^\dagger_{r_\Sigma(t_n)}\tilde{M}_{r_\Sigma(t_n)}\middle|\Psi_\Sigma(t_{n+1})\right\rangle} \\ &= \exp\left(-\frac{\delta Q[\vec{\Sigma},t_n]}{k_{\mathrm{B}}T}\right), \end{aligned} \tag{4.25}$$

and, using the expression of the thermal distribution (Eq. (4.2)), we get

$$\frac{p^{\infty}_{\beta_0}[\epsilon_\Sigma(t_0)]}{p^{\infty}_{\beta_\Sigma(t_N)}[\epsilon_\Sigma(t_N)]} = \exp\left(\frac{\Delta\varepsilon_{\mathrm{q}}[\vec{\Sigma}]-\Delta F[\vec{\Sigma}]}{k_{\mathrm{B}}T}\right). \tag{4.26}$$

The initial and final thermal distributions respectively depend on β_0 and $\beta_\Sigma(t_N)$, which leads to a trajectory-dependent free energy variation

$$\Delta F[\vec{\Sigma}] = k_{\mathrm{B}}T\log\left(\frac{Z(\beta_0)}{Z(\beta_\Sigma(t_N))}\right). \tag{4.27}$$

In the semi-classical regime, the partition function can be approximated by $Z(\beta_\Sigma(t_N)) \simeq Z(\beta_0(t_N))$, so we recover the usual trajectory-independent free energy variation ΔF. Injecting all the above results in Eq. (4.24), we get

$$s_{\mathrm{irr}}[\vec{\Sigma}] = -\log(p_{\mathrm{m}}[\beta_\Sigma(t_N)]) + \frac{\Delta\varepsilon_{\mathrm{q}}[\vec{\Sigma}] - \Delta F[\vec{\Sigma}] - Q[\vec{\Sigma}]}{k_{\mathrm{B}}T} \tag{4.28}$$

Finally, using Eqs. (4.21) and (4.23), we obtain the following expression for the stochastic entropy produced along $\vec{\Sigma}$:

$$s_{\mathrm{irr}}[\vec{\Sigma}] = \sigma[\vec{\Sigma}] + I_{\mathrm{Sh}}[\vec{\Sigma}], \tag{4.29}$$

where $\sigma[\vec{\Sigma}]$ and $I_{\mathrm{Sh}}[\vec{\Sigma}]$ are defined as

$$\sigma[\vec{\Sigma}] := -\frac{\Delta\varepsilon_{\mathrm{m}}[\vec{\Sigma}] + \Delta F[\vec{\Sigma}]}{k_{\mathrm{B}}T}, \tag{4.30}$$

$$I_{\mathrm{Sh}}[\vec{\Sigma}] := -\log(p_{\mathrm{m}}[\beta_\Sigma(t_N)]). \tag{4.31}$$

As shown below, in the semi-classical regime, $\sigma[\vec{\Sigma}]$ can be interpreted as the entropy production along the reduced trajectory of the qubit, giving rise to a reduced JE. On the other hand, $I_{\mathrm{Sh}}[\vec{\Sigma}]$ corresponds to the stochastic increase in entropy of the MO and is involved in a generalized IFT. Next section is dedicated to the study of these two theorems.

4.3 Fluctuation Theorems

4.3.1 Jarzynski Equality

We first focus on the transformation undergone by the qubit which corresponds to Jarzynski's protocol in the semi-classical regime, as mentioned previously: The qubit is driven out of equilibrium by the Hamiltonian $H_{\mathrm{q}}^{\mathrm{eff}}(t) = \hbar\omega(\beta_0(t))\,|e\rangle\langle e|$. We therefore expect the mechanical energy fluctuations to obey the reduced JE

$$\left\langle \exp\left(\frac{\Delta\varepsilon_{\mathrm{m}}[\vec{\Sigma}]}{k_{\mathrm{B}}T}\right)\right\rangle_{\vec{\Sigma}} = \exp\left(-\frac{\Delta F}{k_{\mathrm{B}}T}\right). \tag{4.32}$$

4.3.1.1 Derivation

This equation is derived by starting from the sum over all reversed trajectories of the whole optomechanical system. Using Eq. (4.16), we obtain

$$1 = \sum_{\overleftarrow{\Sigma}} \tilde{P}[\overleftarrow{\Sigma}]$$

$$= \sum_{\overleftarrow{\Sigma}} p_{\mathrm{m}}[\beta_\Sigma(t_N)] p^{\infty}_{\beta_\Sigma(t_N)}[\epsilon_\Sigma(t_N)] \prod_{n=N-1}^{0} \tilde{P}[\Psi_\Sigma(t_n)|\Psi_\Sigma(t_{n+1})]. \tag{4.33}$$

In the semi-classical limit $|\beta_0| \gg g_{\mathrm{m}}/\Omega$, the action of the MO on the qubit is similar to an external operator imposing the evolution of the qubits frequency $\omega(\beta_0(t))$. As a consequence, the reversed jump probability at time t_n (Eq. (4.17)) does not depend on the exact MO state $\beta_\Sigma(t_n)$, but only on $\beta_0(t_n)$, which corresponds to the free MO dynamics. Therefore, we can get rid of the trajectory dependencies in the MO state:

$$\tilde{P}[\Psi_\Sigma(t_n)|\Psi_\Sigma(t_{n+1})] = \tilde{P}[\epsilon_\Sigma(t_n)|\epsilon_\Sigma(t_{n+1})], \tag{4.34}$$

$$p^{\infty}_{\beta_\Sigma(t_N)}[\epsilon_\Sigma(t_N)] = p^{\infty}_{\beta_0(t_N)}[\epsilon_\Sigma(t_N)]. \tag{4.35}$$

Injecting these approximations in Eq. (4.33), we obtain

$$1 = \left(\sum_{\beta_\Sigma(t_N)} p_{\mathrm{m}}[\beta_\Sigma(t_N)]\right) \sum_{\overleftarrow{\epsilon}} p^{\infty}_{\beta_0(t_N)}[\epsilon_\Sigma(t_N)] \prod_{n=N-1}^{0} \tilde{P}[\epsilon_\Sigma(t_n)|\epsilon_\Sigma(t_{n+1})] \tag{4.36}$$

$$= \sum_{\overleftarrow{\epsilon}} p^{\infty}_{\beta_0(t_N)}[\epsilon_\Sigma(t_N)] \prod_{n=0}^{N-1} \tilde{P}[\epsilon_\Sigma(t_n)|\epsilon_\Sigma(t_{n+1})], \tag{4.37}$$

where we have used Eq. (4.13). Moreover, assuming that the temperature in the bath is finite, then all transition probabilities between the qubit's states are non zero and so is the probability $P[\vec{\epsilon}\,]$ of the reduced trajectory of the qubit. Therefore, we can write

$$1 = \sum_{\vec{\epsilon}} P[\vec{\epsilon}\,] \frac{p^{\infty}_{\beta_0(t_N)}[\epsilon_\Sigma(t_N)] \prod_{n=0}^{N-1} \tilde{P}[\epsilon_\Sigma(t_n)|\epsilon_\Sigma(t_{n+1})]}{p^{\infty}_{\beta_0}[\epsilon_\Sigma(t_0)] \prod_{n=0}^{N-1} P[\epsilon_\Sigma(t_{n+1})|\epsilon_\Sigma(t_n)]}. \tag{4.38}$$

Since the trajectory of the MO $\vec{\beta}[\vec{\epsilon}\,]$ is completely determined by the one of the qubit, we can restore the sum over the trajectories $\vec{\Sigma}$ of the whole optomechanical system. Then, from Eqs. (4.26), (4.23) and (4.25), we get

$$1 = \sum_{\vec{\Sigma}} P[\vec{\Sigma}]\exp\left(-\frac{\Delta\varepsilon_{\mathrm{q}}[\vec{\Sigma}] - \Delta F - Q[\vec{\Sigma}]}{k_{\mathrm{B}}T}\right)$$

$$= \left\langle \exp\left(\frac{\Delta\varepsilon_{\mathrm{m}}[\vec{\Sigma}] + \Delta F}{k_{\mathrm{B}}T}\right)\right\rangle_{\vec{\Sigma}}.$$

Finally, in the semi-classical limit, the reduced entropy production $\sigma[\vec{\Sigma}]$ obeys the reduced IFT

$$\left\langle \exp(-\sigma[\vec{\Sigma}]) \right\rangle_{\vec{\Sigma}} = 1, \tag{4.39}$$

analogous to the IFT (2.56) from Chap. 2.

4.3.1.2 Discussion

Equation (4.32) corresponds to the usual JE but where the stochastic work $W[\vec{\Sigma}]$ has been replaced by the mechanical energy variation $\Delta\varepsilon_{\mathrm{m}}[\vec{\Sigma}]$. This is the third key result of this chapter because it suggests a new strategy to measure JE in a quantum open system. Instead of monitoring the complete trajectory of the system to reconstruct the stochastic work, we propose to simply measure the stochastic mechanical energy at the beginning and at the end of the transformation. This can be achieved with time resolved measurements of the mechanical amplitude through optical deflection techniques [24, 25]. Therefore, the mechanical states $|\beta_{\Sigma}(t_N)\rangle$ have to be distinguishable, which requires the ultra-strong coupling regime. As mentioned in Chap. 3 (See Table 3.1), this regime is experimentally reachable. This strategy is very different from former proposals to probe JE in a quantum open system which used bath engineering techniques [16, 17] or fine thermometry [15] to monitor heat exchanges.

We have simulated the reduced JE using experimentally realistic parameters. The results are displayed in Fig. 4.2. The plots in this figure and in Fig. 4.3 were obtained by approximating the average value of the plotted quantity $A[\vec{\Sigma}]$ by

$$\langle A\rangle_{\vec{\Sigma}} \simeq \frac{1}{N_{\mathrm{traj}}}\sum_{i=1}^{N_{\mathrm{traj}}} A[\vec{\Sigma}_i], \tag{4.40}$$

where $N_{\mathrm{traj}} = 5\cdot 10^6$ is the number of numerically generated trajectories and $\vec{\Sigma}_i$ denotes the i-th simulated trajectory. $\sigma[\vec{\Sigma}_i]$ was computed using Eq. (4.30), i.e. from the complex mechanical amplitudes β_0 and $\beta^i_{\Sigma}(t_N)$. As expected, JE is verified in the semi-classical limit (Fig. 4.2a), in which we have checked that the MO action is equivalent to the one of a classical external operator imposing the modulation $\omega(\beta_0(t))$ to the qubit's transition frequency (Fig. 4.2b). Reciprocally, the Markovian approximation for the reduced trajectory of the qubit and JE break down in the regime $(g_{\mathrm{m}}/\Omega)/|\beta_0| \geq 10^{-2}$. Therefore, in the following we only consider parameters such that $(g_{\mathrm{m}}/\Omega)/|\beta_0| < 10^{-2}$.

Up to now, we have assumed that the mechanical states could be measured with an infinite precision. To take into account both the quantum uncertainties and the experimental sources of imprecision, we assume that the measured complex amplitude β^{M} corresponds to the mechanical amplitude with a finite precision $\delta\beta$. This finite precision is quantified by the mutual information between the probability distribution

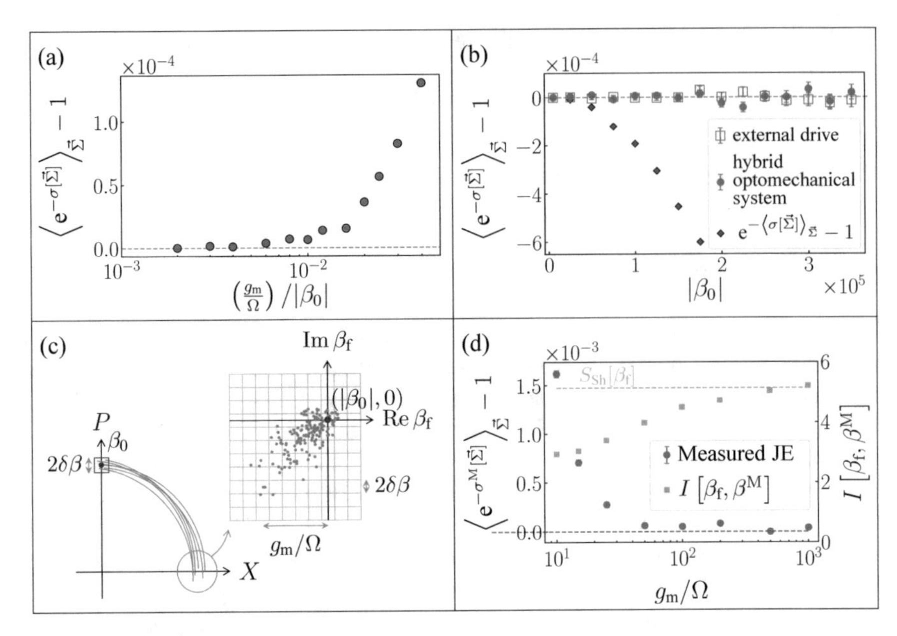

Fig. 4.2 Jarzynski equality for the qubit. Parameters: $T = 80$ K, $\hbar\omega_0 = 1.2k_BT$, $\gamma/\Omega = 5$. **a** Deviation from JE as a function of $\left(\frac{g_m}{\Omega}\right)/|\beta_0|$ ($\Omega/2\pi = 100$ kHz, $|\beta_0| = 5000$). The points were computed by increasing the opto-mechanical coupling strength $g_m/2\pi$ from 1 MHz to 20 MHz, keeping the other parameters constant. **b** Deviation from JE as a function of $|\beta_0|$ with $g_m/\Omega = 10$ and $\Omega/2\pi = 100$ kHz. Red squares: Case of a classical external drive imposing the qubit frequency modulation $\omega(\beta_0(t))$. Blue dots: Eq. (4.32). Green diamonds: $\exp\left(-\left\langle\sigma[\vec{\Sigma}]\right\rangle_{\vec{\Sigma}}\right) - 1$. These green points demonstrate that JE is not trivially reached because the considered transformations are reversible. **c** Error model. The initial mechanical state is randomly drawn in the square of width $2\delta\beta$ around the targeted state $|\beta_0\rangle$ then the MO evolves for a quarter period. The final states $|\beta_f\rangle$ obtained for different trajectories are plotted in the inset. The grid represent precision of the measurement device. When measuring $|\beta_f\rangle$, instead of obtaining the exact mechanical amplitude, we obtain β^M which is the center of the grid cell containing $|\beta_f\rangle$. **d** Impact of finite precision readout of the mechanical amplitude for $\delta\beta = 2$ and $\Omega/2\pi = 1$ kHz. $2g_m|\beta_0|$ was kept constant ($2g_m|\beta_0|/2\pi = 600$ GHz) while increasing g_m, such that each point corresponds to the same mean reduced entropy production $\langle\sigma[\vec{\Sigma}]\rangle_{\vec{\Sigma}}$. Left axis, blue dots: Deviation from measured JE. Right axis, orange squares: Mutual information $I[\beta_f, \beta^M]$. Orange dashed line: Shannon entropy of the final distribution of mechanical states $S_{Sh}[\beta_f]$. The error bars in **b** and **d** represent the standard error of the mean $\sigma/\sqrt{N_{traj}}$, where σ is the standard deviation and $N_{traj} = 5 \cdot 10^6$ is the number of numerically generated trajectories. See Appendix A.1 for more details about the numerical simulations

of the final mechanical states $p_m[\beta_f]$, introduced in Sect. 4.1.1, and the probability distribution of the measured states $p_m[\beta^M]$, defined by

$$I[\beta_f, \beta^M] := \sum_{\beta_f, \beta^M} p(\beta_f, \beta^M) \log\left(\frac{p(\beta_f, \beta^M)}{p_m[\beta_f] p_m[\beta^M]}\right). \tag{4.41}$$

$p(\beta_\mathrm{f}, \beta^\mathrm{M})$ is the joint probability to measure β^M while the mechanical amplitude is β_f. For a perfect measurement, the mutual information is equal to the Shannon entropy

$$S_\mathrm{sh}[\beta_\mathrm{f}] = -\sum_{\beta_\mathrm{f}} p_\mathrm{m}[\beta_\mathrm{f}]\log(p_\mathrm{m}[\beta_\mathrm{f}]) \tag{4.42}$$

that characterizes the final distribution of mechanical states. Conversely, the mutual information vanishes when the two distributions are totally uncorrelated.

To be more specific about the finite precision protocol, we first consider that the preparation of the initial MO state is not perfect. Instead of starting from exactly $|\beta_0\rangle$, the mechanical trajectories start from $|\beta_\Sigma(t_0)\rangle$ with $\beta_\Sigma(t_0)$ uniformly distributed in a square of width $2\delta\beta$, centered on β_0 (See Fig. 4.2c). Then, we let the optomechanical system evolve for a mechanical quarter period and measure the final state. The measuring apparatus has a finite precision, modeled by a grid of cell width $2\delta\beta$ in the phase plane $(\mathrm{Re}\beta_\mathrm{f}, \mathrm{Im}\beta_\mathrm{f})$. Instead of obtaining the exact value of $\beta_\Sigma(t_N)$, we get $\beta^\mathrm{M}_\Sigma(t_N)$, the center of the grid cell in which $\beta_\Sigma(t_N)$ is. The value used to compute the thermodynamic quantities are not the exact $\beta_\Sigma(t_0)$ and $\beta_\Sigma(t_N)$ but $\beta_0^\mathrm{M} = \beta_0$ and $\beta^\mathrm{M}_\Sigma(t_N)$. As a result, the measured work and reduced entropy production reads

$$W^\mathrm{M}[\vec{\Sigma}] = -\Delta\varepsilon^\mathrm{M}_\mathrm{m}[\vec{\Sigma}] = \hbar\Omega\left(|\beta_0^\mathrm{M}|^2 - |\beta^\mathrm{M}_\Sigma(t_N)|^2\right), \tag{4.43}$$

$$\sigma^\mathrm{M}[\vec{\Sigma}] = \frac{W^\mathrm{M}[\vec{\Sigma}] - \Delta F}{k_\mathrm{B}T} \tag{4.44}$$

The deviation from JE for the measured work $W^\mathrm{M}[\vec{\Sigma}]$ and the mutual information $I[\beta_\mathrm{f}, \beta^\mathrm{M}]$ are plotted in Fig. 4.2d as a function of g_m/Ω. For the numerical simulations, we have chosen the measurement precision $\delta\beta = 2$, which is a reachable experimental value [24, 25]. For small values of g_m/Ω, the spread of the final mechanical states is barely larger than the measurement precision, leading to a poor ability to distinguish the mechanical states and therefore to measure work. As a consequence, the mutual information is much smaller than the Shannon entropy and $W^\mathrm{M}[\vec{\Sigma}]$ does not verify JE. Conversely, increasing the coupling ratio g_m/Ω increases the spread of the final mechanical states. Therefore, the amount of information extracted during the measurement increases making $I[\beta_\mathrm{f}, \beta^\mathrm{M}]$ converge towards $S_\mathrm{sh}[\beta_\mathrm{f}]$. JE equality is recovered for $g_\mathrm{m}/\Omega \sim 50$ despite the finite precision. Such high optomechanical coupling ratios are within experimental reach, for instance by engineering lower mechanical frequency or by changing the geometry of the MO [26].

4.3.2 *Generalized Integral Fluctuation Theorem*

Finally, we consider the hybrid optomechanical system as the thermodynamic system. The entropy production for the whole system $s_\mathrm{irr}[\vec{\Sigma}]$ (Eq. (4.29)) obeys the generalized IFT

$$\left\langle \exp(-s_{\mathrm{irr}}[\vec{\Sigma}]) \right\rangle_{\vec{\Sigma}} = 1 - \lambda. \tag{4.45}$$

As in Refs. [27–30], we have defined the parameter $\lambda \in [0, 1]$ as

$$\sum_{\overleftarrow{\Sigma}} \tilde{P}[\overleftarrow{\Sigma}] = 1 - \lambda. \tag{4.46}$$

4.3.2.1 Derivation

To derive this theorem, we start from the sum over all reversed trajectories, making appear the ratio $\tilde{P}[\overleftarrow{\Sigma}]/P[\vec{\Sigma}]$. Therefore, we need to ensure that $P[\vec{\Sigma}] \neq 0$, which requires to split the ensemble of the reversed trajectories into the set $\Sigma_{\mathrm{d}} = \{\tilde{P}[\overleftarrow{\Sigma}] | P[\vec{\Sigma}] \neq 0\}$ of reversed trajectories with a direct counterpart and the set without:

$$1 = \sum_{\overleftarrow{\Sigma}} \tilde{P}[\overleftarrow{\Sigma}] = \sum_{\overleftarrow{\Sigma} \in \Sigma_{\mathrm{d}}} P[\vec{\Sigma}] \frac{\tilde{P}[\overleftarrow{\Sigma}]}{P[\vec{\Sigma}]} + \sum_{\overleftarrow{\Sigma} \notin \Sigma_{\mathrm{d}}} \tilde{P}[\overleftarrow{\Sigma}]. \tag{4.47}$$

Only the reversed trajectories $\overleftarrow{\Sigma} = \{|\tilde{\epsilon}_{\Sigma}(t_n), \tilde{\beta}_{\Sigma}(t_n)\rangle\}_{n=N}^{0}$ such that $\tilde{\beta}_{\Sigma}(t_0) = \beta_0$ verify $P[\vec{\Sigma}] \neq 0$. Figures 4.3a and b give examples of both kinds of trajectories. Denoting

$$\lambda = \sum_{\overleftarrow{\Sigma} \notin \Sigma_{\mathrm{d}}} \tilde{P}[\overleftarrow{\Sigma}] \tag{4.48}$$

and using Eqs. (4.9) and (4.16) we obtain:

$$\begin{aligned} 1 &= \sum_{\vec{\Sigma}} \left(P[\vec{\Sigma}] p_{\mathrm{m}}[\beta_{\Sigma}(t_N)] \frac{p^{\infty}_{\beta_{\Sigma}(t_N)}[\epsilon_{\Sigma}(t_N)]}{p^{\infty}_{\beta_0}[\epsilon_{\Sigma}(t_0)]} \frac{\prod_{n=1}^{N} \tilde{P}[\Psi_{\Sigma}(t_{n-1})|\Psi_{\Sigma}(t_n)]}{\prod_{n=1}^{N} P[\Psi_{\Sigma}(t_n)|\Psi_{\Sigma}(t_{n-1})]} \right) + \lambda \\ &= \sum_{\vec{\Sigma}} P[\vec{\Sigma}] \exp\left(-I_{\mathrm{Sh}}[\vec{\Sigma}] - \frac{\Delta\varepsilon_{\mathrm{q}}[\vec{\Sigma}] - \Delta F[\vec{\Sigma}] - Q[\vec{\Sigma}]}{k_{\mathrm{B}} T} \right) + \lambda \\ &= \left\langle \exp(-(\sigma[\vec{\Sigma}] + I_{\mathrm{Sh}}[\vec{\Sigma}])) \right\rangle_{\vec{\Sigma}} + \lambda. \end{aligned} \tag{4.49}$$

Thus, we have derived Eq. (4.45).

4.3.2.2 Discussion

$\lambda > 0$ signals the existence of reversed trajectories without a direct counterpart and quantifies absolute irreversibility [27]. From Eq. (4.45) and the convexity of the exponential, absolute irreversibility clearly characterizes transformations associated to a strictly positive entropy production, as stated in Chap. 2. This is the case for

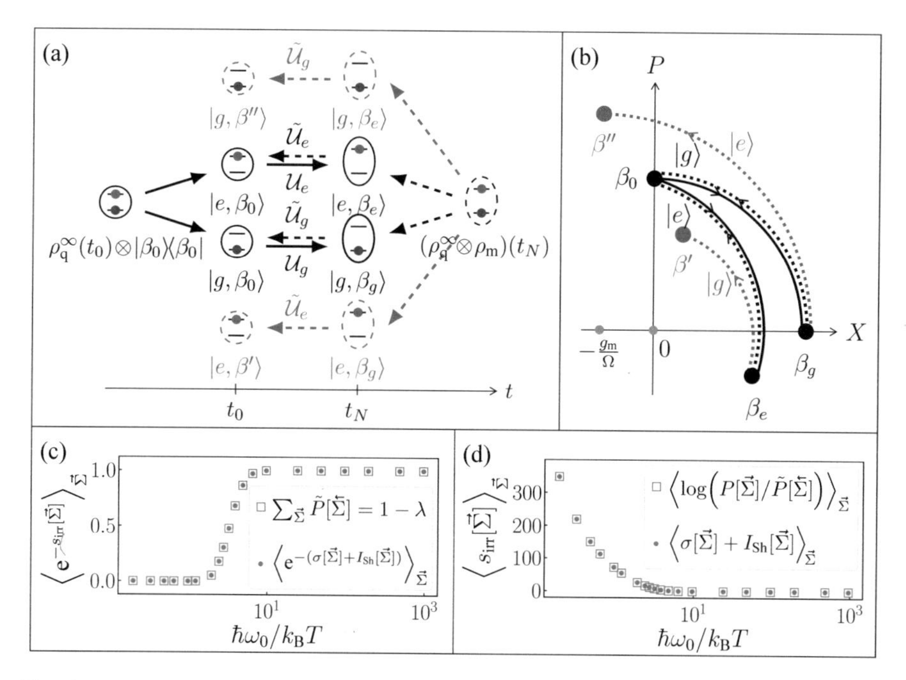

Fig. 4.3 Illustration of absolute irreversibility with trajectories of **a** the qubit and **b** the MO. The solid (resp. dashed) arrows correspond to the direct (resp. reversed) protocol. For the sake of simplicity, only the trajectories without any jump are represented. β_ϵ is the final state of the MO after the direct protocol when the qubit is in state $|\epsilon\rangle$, with $\epsilon = e, g$. The expressions of the MO evolution operators are: $\mathcal{U}_\epsilon(t) = \exp(-\mathrm{i}t H_\mathrm{m}^\epsilon)$ and $\tilde{\mathcal{U}}_\epsilon(t) = \mathcal{U}_\epsilon^\dagger(t)$. The reversed trajectories that do not have a direct counterpart are plotted in red and the corresponding qubit states with dashed lines. The final MO states for these trajectories are $|\beta''\rangle = \tilde{\mathcal{U}}_g(t_N)\,|\beta_e\rangle$ and $|\beta'\rangle = \tilde{\mathcal{U}}_e(t_N)\,|\beta_g\rangle$, where $\beta'' \neq \beta_0$ and $\beta' \neq \beta_0$. $\rho_\mathrm{q}^\infty(t)$ (resp. $\rho_\mathrm{m}(t)$) is the qubit thermal state (resp. the MO average state) at time t. **c** Integral fluctuation theorem and **d** mean entropy production for the complete optomechanical system. Parameters: $\omega_0/2\pi = 2$ THz (amounts to $\hbar\omega_0/k_\mathrm{B}T = 1.2$ for $T = 80$ K used in Fig. 4.2), $\Omega/2\pi = 100$ kHz, $\gamma/\Omega = 5$, $g_\mathrm{m}/\Omega = 10$ and $|\beta_0| = 5000$. In both cases, two different expressions were used. The blue dots are computed using the final distribution of mechanical states $\{|\beta_\Sigma(t_N)\rangle\}$ and mimic an experiment. The red squares involve the probability of the reversed trajectory, which can only be the result of a theoretical treatment. See Appendix A.1 for more details about the numerical simulations

the transformation under study here, which is the relaxation of the optomechanical system towards a thermal equilibrium state. Indeed, such transformation is never reversible, unless for $T = 0$, as confirmed by Fig. 4.3d.

Both sides of the IFT (Eq. (4.45)) are plotted in Fig. 4.3c as a function of the bath temperature T, the left hand side (blue dots) was computed with Eq. (4.29) and the right hand side (red squares) with Eq. (4.46) from the probabilities of the reversed trajectories. One value of $\beta_\Sigma(t_N)$ can be generated by a single direct trajectory $\vec{\Sigma}$, so using the equality $p_\mathrm{m}[\beta_\Sigma(t_N)] = P[\vec{\Sigma}]$, we obtain

$$\sum_{\vec{\Sigma}} \tilde{P}[\overleftarrow{\Sigma}] = \sum_{\vec{\Sigma}} p^{\infty}_{\beta_\Sigma(t_N)}[\epsilon_\Sigma(t_N)] p_{\rm m}[\beta_\Sigma(t_N)] \prod_{n=0}^{N-1} \tilde{P}[\Psi_\Sigma(t_n)|\Psi_\Sigma(t_{n+1})]$$

$$= \left\langle p^{\infty}_{\beta_\Sigma(t_N)}[\epsilon_\Sigma(t_N)] \prod_{n=1}^{N} \tilde{P}[\Psi_\Sigma(t_{n-1})|\Psi_\Sigma(t_n)] \right\rangle_{\vec{\Sigma}} , \tag{4.50}$$

which is then approximated by Eq. (4.40). The average entropy production is plotted in Fig. 4.3d as a function of the bath temperature T. It was numerically computed with two methods: by taking the average of Eq. (4.29) (blue dots), approximated by Eq. (4.40), and directly from the definition Eq. (2.50) (red squares), i.e from the probabilities of the trajectories:

$$\left\langle \Delta_{\rm i} s[\vec{\Sigma}] \right\rangle_{\vec{\Sigma}} = \left\langle \log \left(\frac{P[\vec{\Sigma}]}{\tilde{P}[\overleftarrow{\Sigma}]} \right) \right\rangle_{\vec{\Sigma}} \tag{4.51}$$

$$= \left\langle -\log \left(p^{\infty}_{\beta_\Sigma(t_N)}[\epsilon_\Sigma(t_N)] \prod_{n=0}^{N-1} \tilde{P}[\Psi_\Sigma(t_n)|\Psi_\Sigma(t_{n+1})] \right) \right\rangle_{\vec{\Sigma}} , \tag{4.52}$$

where we have used the expression (4.16) of the probability of the reversed trajectory and $p_{\rm m}[\beta_\Sigma(t_N)] = P[\vec{\Sigma}]$. Figures 4.3c and d show that both sets of points match, however only the blue dots, obtained from the final distribution of mechanical states $\{|\beta_\Sigma(t_N)\rangle\}_{\vec{\Sigma}}$ are experimentally accessible. In the limit $\hbar\omega_0 \gg k_{\rm B}T$, the bath contains no photons at the qubit's frequency, therefore there is single reversible trajectory characterized by a null entropy production and $\lambda \to 0$. In the opposite limit, $k_{\rm B}T \gg \hbar\omega_0$, most reversed trajectories have no direct counterpart. Indeed, a given mechanical state $|\beta_{\rm f}\rangle$ of the final distribution can only be reached by a single direct trajectory, while it provides a starting point for a large number of reversed trajectories. As a consequence, a mean entropy is produced while $\lambda \to 1$.

References [23, 27, 30] show that absolute irreversibility can also appear in IFTs characterizing the entropy produced by a measurement process. In particular, $\lambda > 0$ can signal a perfect information extraction. This typically corresponds to the present situation which describes the creation of classical correlations between the qubit reduced trajectory $\vec{\epsilon}$ and the distributions of final mechanical states $\{|\beta_\Sigma(t_N)\rangle\}_{\vec{\Sigma}}$. Interestingly, the two fluctuation theorems (4.32) and (4.45) are thus deeply related. To be experimentally checked, Eq. (4.32) requires the MO to behave as a perfect quantum work meter, which is signaled by absolute irreversibility in Eq. (4.45). Therefore absolute irreversibility is constitutive of the protocol, and a witness of its success.

4.4 Summary

We studied the hybrid optomechanical system in the quantum trajectory picture. We obtained the stochastic evolution of the system by unraveling the master equation derived in the previous chapter. First, we showed that when the optomechanical system is prepared in a tensor product of an energy eigenstate of the qubit and a coherent mechanical state, it remains in a state of the same form all along the trajectory. This result allowed us to define without ambiguity the qubit's energy and the mechanical energy.

Secondly, we defined the thermodynamic quantities at the single trajectory level. Going one step further than in Chap. 3, we evidenced that work fluctuations equal the mechanical energy fluctuations, which are measurable in the ultra-strong coupling regime. Therefore, stochastic work exchanges can be directly obtained by measuring the energy of the battery at the beginning and at the end of the thermodynamic transformation. This method of work measurement based on the direct readout of work exchanges within an autonomous machine offers a promising alternative to proposals involving system and/or bath monitoring. Based on this result, we proposed a new protocol to measure stochastic entropy production and the thermodynamic time arrow in a quantum open system.

Finally, we investigated fluctuations theorems both in the perspective of the qubit and of the whole optomechanical system. When the thermodynamic system is the qubit, the transformation is an out-of-equilibrium driving and, in the Markovian limit, the reduced entropy production along the qubit's trajectory obeys Jarzynski equality. We then evidenced that our protocol can be used to experimentally probe this fluctuation theorem in state-of-the-art optomechanical devices. In the perspective of the whole optomechanical system, the transformation is a relaxation toward equilibrium which is therefore strictly irreversible. We showed that the total entropy production obeys a generalized integral fluctuation theorem, shedding new light on absolute irreversibility, which quantifies information extraction within the quantum work meter and therefore signals the success of the protocol.

Generalizing our formalism to other kind of autonomous machines would open ways to investigate genuinely quantum situations where a battery coherently drives a quantum open system into coherent superpositions. Such situations are especially appealing for quantum thermodynamics since they lead to entropy production and energetic fluctuations of quantum nature [20, 31], related to the erasure of quantum coherences [3, 4].

References

1. Mancino L, Cavina V, De Pasquale A, Sbroscia M, Booth RI, Roccia E, Gianani I, Giovannetti V, Barbieri M (2018) Geometrical bounds on irreversibility in open quantum systems. Phys Rev Lett 121(16):160 602. https://doi.org/10.1103/PhysRevLett.121.160602
2. Mancino L, Sbroscia M, Roccia E, Gianani I, Somma F, Mataloni P, Paternostro M, Barbieri M (2018) The entropic cost of quantum generalized measurements. Npj Quantum Inf 4(1):20. https://doi.org/10.1038/s41534-018-0069-z
3. Santos JP, Céleri LC, Landi GT, Paternostro M (2019) The role of quantum coherence in non-equilibrium entropy production. Npj Quantum Inf 5(1). https://doi.org/10.1038/s41534-019-0138-y
4. Francica G, Goold J, Plastina F (2019) Role of coherence in the nonequilibrium thermodynamics of quantum systems. Phys Rev E 99(4):042 105. https://doi.org/10.1103/PhysRevE.99.042105
5. Talkner P, Lutz E, Hänggi P (2007) Fluctuation theorems: work is not an observable. Phys Rev E 75(5):050 102. https://doi.org/10.1103/PhysRevE.75.050102
6. Esposito M, Harbola U, Mukamel S (2009) Nonequilibrium fluctuations, fluctuation theorems, and counting statistics in quantum systems. Rev Mod Phys 81(4):1665–1702. https://doi.org/10.1103/RevModPhys.81.1665
7. Talkner P, Hänggi P (2016) Aspects of quantum work. Phys Rev E 93(2):022 131. https://doi.org/10.1103/PhysRevE.93.022131
8. Bäumer E, Lostaglio M, Perarnau-Llobet M, Sampaio R (2018) Fluctuating work in coherent quantum systems: proposals and limitations. In: Binder F, Correa LA, Gogolin C, Anders J, Adesso G (eds) Thermodynamics in the quantum regime. Springer, Cham
9. Engel A, Nolte R (2007) Jarzynski equation for a simple quantum system: comparing two definitions of work. Europhys Lett 79(1):10 003. https://doi.org/10.1209/0295-5075/79/10003
10. Campisi M, Hänggi P, Talkner P (2011) Colloquium: quantum fluctuation relations: foundations and applications. Rev Mod Phys 83(3):771–791. https://doi.org/10.1103/RevModPhys.83.771
11. An S, Zhang J-N, Um M, Lv D, Lu Y, Zhang J, Yin Z-Q, Quan HT, Kim K (2015) Experimental test of the quantum Jarzynski equality with a trapped-ion system. Nat Phys 11(2):193. https://doi.org/10.1038/nphys3197
12. Xiong TP, Yan LL, Zhou F, Rehan K, Liang DF, Chen L, Yang WL, Ma ZH, Feng M, Vedral V (2018) Experimental verification of a Jarzynski-related information-theoretic equality by a single trapped ion. Phys Rev Lett 120(1):010 601. https://doi.org/10.1103/PhysRevLett.120.010601
13. Cerisola F, Margalit Y, Machluf S, Roncaglia AJ, Paz JP, Folman R (2017) Using a quantum work meter to test non-equilibrium fluctuation theorems. Nat Commun 8(1):1241. https://doi.org/10.1038/s41467-017-01308-7
14. Batalhão TB, Souza AM, Mazzola L, Auccaise R, Sarthour RS, Oliveira IS, Goold J, De Chiara G, Paternostro M, Serra RM (2014) Experimental reconstruction of work distribution and study of fluctuation relations in a closed quantum system. Phys Rev Lett 113(14):140 601. https://doi.org/10.1103/PhysRevLett.113.140601
15. Pekola JP, Solinas P, Shnirman A, Averin DV (2013) Calorimetric measurement of work in a quantum system. New J Phys 15(11):115 006. https://doi.org/10.1088/1367-2630/15/11/115006
16. Horowitz JM (2012) Quantum-trajectory approach to the stochastic thermodynamics of a forced harmonic oscillator. Phys Rev E 85(031):110. https://doi.org/10.1103/PhysRevE.85.031110
17. Elouard C, Bernardes NK, Carvalho ARR, Santos MF, Auffèves A (2017) Probing quantum fluctuation theorems in engineered reservoirs. New J Phys 19(10):103011. https://doi.org/10.1088/1367-2630/aa7fa2
18. Monsel J, Elouard C, Auffèves A (2018) An autonomous quantum machine to measure the thermodynamic arrow of time. Npj Quantum Inf 4(1):59. https://doi.org/10.1038/s41534-018-0109-8

19. Haroche S, Raimond J-M (2006) Exploring the quantum: atoms, cavities, and photons. Oxford University Press, Oxford
20. Elouard C, Herrera-Martí DA, Clusel M, Auffèves A (2017) The role of quantum measurement in stochastic thermodynamics. Npj Quantum Inf 3(1):9. https://doi.org/10.1038/s41534-017-0008-4
21. Crooks GE (2008) Quantum operation time reversal. Phys Rev A 77(3):034 101. https://doi.org/10.1103/PhysRevA.77.034101
22. Manzano G, Horowitz JM, Parrondo JMR (2018) Quantum fluctuation theorems for arbitrary environments: adiabatic and nonadiabatic entropy production. Phys Rev X 8(3):031 037. https://doi.org/10.1103/PhysRevX.8.031037
23. Manikandan SK, Jordan AN (2019) Time reversal symmetry of generalized quantum measurements with past and future boundary conditions. Quantum Stud Math Found 6(2):241–268. https://doi.org/10.1007/s40509-019-00182-w
24. Sanii B, Ashby PD (2010) High sensitivity deflection detection of nanowires. Phys Rev Lett 104(14):147 203. https://doi.org/10.1103/PhysRevLett.104.147203
25. Mercier de Lépinay L, Pigeau B, Besga B, Vincent P, Poncharal P, Arcizet O (2017) A universal and ultrasensitive vectorial nanomechanical sensor for imaging 2D force fields. Nat Nanotechnol 12(2):156–162. https://doi.org/10.1038/nnano.2016.193
26. Yeo I, de Assis P-L, Gloppe A, Dupont-Ferrier E, Verlot P, Malik NS, Dupuy E, Claudon J, Gérard J-M, Auffèves A, Nogues G, Seidelin S, Poizat J-P, Arcizet O, Richard M (2014) Supplementary information for "strain-mediated coupling in a quantum dot-mechanical oscillator hybrid system". Nat Nanotech 9:106–110. https://doi.org/10.1038/nnano.2013.274
27. Murashita Y, Funo K, Ueda M (2014) Nonequilibrium equalities in absolutely irreversible processes. Phys Rev E 90(4):042 110. https://doi.org/10.1103/PhysRevE.90.042110
28. Funo K, Murashita Y, Ueda M (2015) Quantum nonequilibrium equalities with absolute irreversibility. New J Phys 17(7):075 005. https://doi.org/10.1088/1367-2630/17/7/075005
29. Murashita Y, Gong Z, Ashida Y, Ueda M (2017) Fluctuation theorems in feedbackcontrolled open quantum systems: quantum coherence and absolute irreversibility. Phys Rev A 96(4):043 840. https://doi.org/10.1103/PhysRevA.96.043840
30. Masuyama Y, Funo K, Murashita Y, Noguchi A, Kono S, Tabuchi Y, Yamazaki R, Ueda M, Nakamura Y (2018) Information-to-work conversion by Maxwell's demon in a superconducting circuit quantum electrodynamical system. Nat Commun 9(1):1291. https://doi.org/10.1038/s41467-018-03686-y
31. Elouard C, Herrera-Martí D, Huard B, Auffèves A (2017) Extracting work from quantum measurement in Maxwell's demon engines. Phys Rev Lett 118(26):260 603. https://doi.org/10.1103/PhysRevLett.118.260603

Chapter 5
Optomechanical Energy Conversion

In this chapter, we analyze the hybrid optomechanical system as a reversible thermal machine, like three-level masers [1], and, like the two level maser from Ref. [2], this machine operates autonomously. Unlike in Chaps. 3 and 4, a laser is shone on the qubit. We consider the saturated regime of the Rabi oscillations, so that the coupling between the laser and the qubit is incoherent and we can identify the laser with the hot bath. The cold bath is the electromagnetic reservoir at zero temperature coupled to the qubit.

19th-century thermal machines are called reversible because they can operate as engines or refrigerators, as illustrated in Fig. 5.1a and b. In the former operating mode, heat flows from the hot bath to the cold bath through the system that provides work to the battery. In the latter operating mode, all energy flows are reversed. The battery provides work to the system which makes the heat flow from the cold bath to the hot bath. What we mean by reversible in the case of the hybrid optomechanical system is different from this usual definition, because the heat always flows from the hot bath to the cold bath, only the work flow is reversed.

The optomechanical coupling results in a modulation of the frequency of the qubit, making it enter in and out of resonance with the laser which enables optomechanical energy conversion. When the laser is blue-detuned (Fig. 5.1c), the qubit receives energy from the hot bath, in the form of high energy photons, gives part of it to the MO as work and dumps the remaining energy inside the cold bath, in the form of lower energy photons. Therefore optical energy is converted into mechanical energy and the hybrid optomechanical system operates as an engine. Conversely, when the laser is red-detuned (Fig. 5.1d), it provides low energy photons to the qubit, which emits higher energy photons in the cold bath, the energy difference is provided by the MO in the form of work. Therefore, the optomechanical system operates as an accelerator, facilitating heat flow from the hot bath to the cold bath and the direction of the optomechanical energy conversion is reversed.

J. Monsel, *Quantum Thermodynamics and Optomechanics*, Springer Theses,
https://doi.org/10.1007/978-3-030-54971-8_5

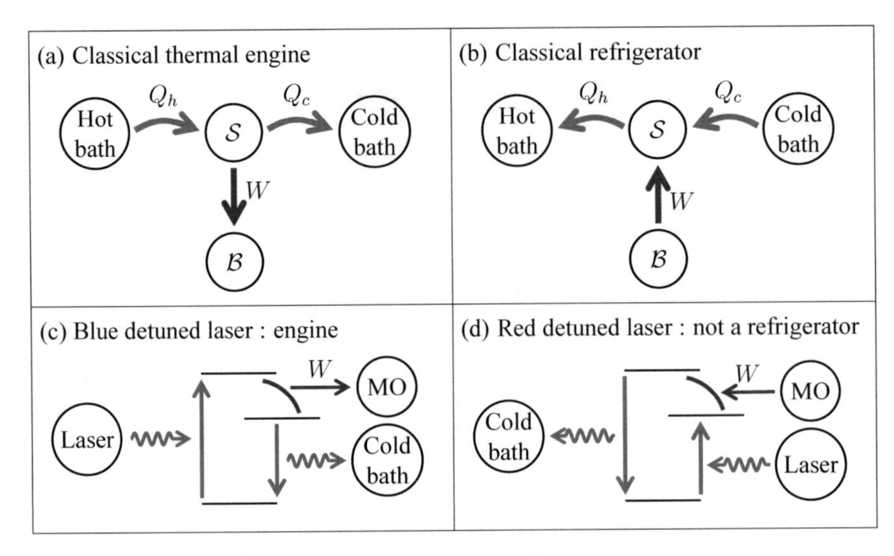

Fig. 5.1 **a** and **b** Classical reversible thermal machine: **a** When heat flows from the hot bath to the cold bath through the system, work is provided to the battery. The thermal machine operates as an engine. **b** Reciprocally, when the battery provides work to the system, the heat flow is reversed and goes from the cold bath to the hot bath. The thermal machine operates as a refrigerator. **c** and **d** Hybrid optomechanical system analyzed as a reversible thermal machine: **c** When the laser is blue-detuned, it provides high energy photons to the qubit, which emits lower energy photons in the cold bath, the energy difference is given to the MO in the form of work. The optomechanical system operates as an engine. **c** When the laser is red-detuned, it provides low energy photons to the qubit, which emits higher energy photons in the cold bath, the energy difference is provided by the MO in the form of work. The optomechanical system operates as an accelerator, facilitating heat flow from the hot bath to the cold bath

We look at longer time scales than in the previous chapters and, therefore, we take into account the environment of the MO. We demonstrate that, with a blue detuning, a coherent phonon state can be built starting from thermal fluctuations, which exhibits laser-like signatures. There have already been a few proposals to make phonon lasers using cavity optomechanics [3–5] and hybrid optomechanical systems [6]. In particular, Ref. [6] proposes to amplify the mechanical motion by driving the qubit with a laser, but in this chapter we enter more into the details of the conversion mechanism using a quantum trajectory approach. Unlike in Ref. [6], we also investigates the other direction of energy conversion, when the laser is red-detuned. We evidence that the average phonon number in the MO can be reduced below the thermal level.

In this chapter, we first sketch the energy conversion principle. Secondly, we describe the dynamics of the hybrid optomechanical system: We derive a master equation that takes into account the bath of the MO, unravel it into stochastic trajectories and finally include an effective modeling of the interaction between the qubit and the laser. Then, we present a simple coarse-grained model of the evolution of the phonon number in the MO, allowing us to estimate the steady-state phonon number and to identify interesting regimes. Finally, we use numerically generated quantum trajectories of the MO to characterize both energy conversion processes.

5.1 Principle of Optomechanical Energy Conversion

If the MO is in a coherent state, the optomechanical coupling results in a sinusoidal modulation of the transition frequency of the qubit (See Chap. 3, Eq. (3.41)). This modulation can be used to convert optical energy into mechanical energy. We first present an ideal optomechanical energy converter and then we explain the principle of an autonomous one.

5.1.1 *Ideal Converter*

Assuming the hybrid optomechanical system is isolated and has been prepared in the state $|\epsilon_0, \beta_0\rangle$ at time t_0, with $\epsilon_0 \in \{e, g\}$ and $|\beta_0\rangle$ a coherent state of the MO, then at time t, the state of the system is

$$\left|\epsilon_0, \beta_{\epsilon_0}(t)\right\rangle = \left|\epsilon_0, \left(\beta_0 + \frac{g_\mathrm{m}}{\Omega}\delta_{\epsilon_0,e}\right)\mathrm{e}^{-\mathrm{i}\Omega(t-t_0)} - \frac{g_\mathrm{m}}{\Omega}\delta_{\epsilon_0,e}\right\rangle. \tag{5.1}$$

Then, the ideal energy conversion protocol, schematized in Fig. 5.2, is the following: When the qubit's frequency reaches its maximum, the qubit is excited, i.e. put in the $|e\rangle$ state, so that the rest position of the MO is shifted by $-2x_{\mathrm{zpf}}g_\mathrm{m}/\Omega$. Then, we let the system evolve for a half mechanical period, according to Eq. (5.1). When the qubit's frequency reaches its minimum, the qubit is de-excited, i.e. put in the $|g\rangle$ state, so the rest position of the MO goes back to 0. Again, we let the system evolve for a half period, then we start again. At each half-period, the radius of the half-circle described by the MO in the phase space (X, P) increases, as shown in Fig. 5.2b.

We can physically analyze this evolution in term of work exchanges using Eq. (3.56): $W = \hbar\Delta\omega P_e$. When the qubit's frequency decreases, $\Delta\omega < 0$ and $P_e = 1$ so an amount of work $\hbar|\Delta\omega|$ is provided to the MO. On the contrary, when the qubit's frequency increases $P_e = 0$, so no work is provided by the MO. Therefore, every mechanical period, the mechanical energy increases by $\hbar|\Delta\omega|$, which translates into an increase in the mechanical amplitude. This ideal energy conversion protocol is reversible. If we excite the qubit when its frequency is minimal and de-excite it when it is maximal, then $P_e = 1$ while the qubit's frequency increases. Therefore, the MO provides an amount of work $\hbar|\Delta\omega|$ every mechanical period, which results in a decrease in the mechanical amplitude.

This protocol could be realized using for instance π-pulses with a Rabi frequency $g \gg \Omega$ so that they would be instantaneous compared to the mechanical evolution. An alternative described in Ref. [7], analyzes this protocol as a heat engine, using

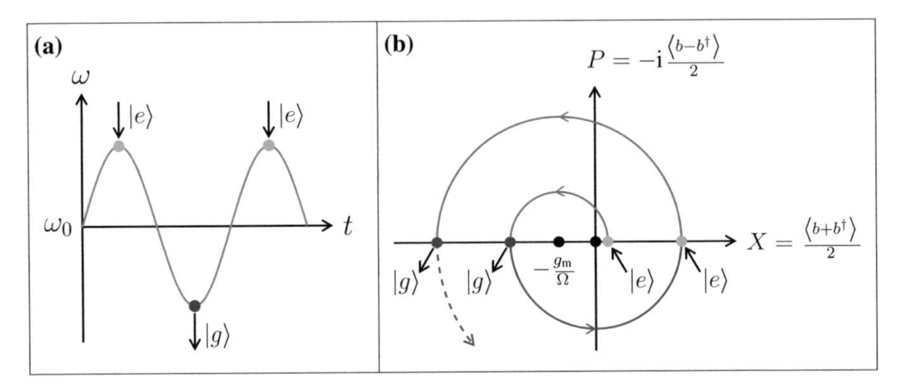

Fig. 5.2 Ideal energy converter. **a** Time evolution of the transition frequency of the qubit. **b** Mechanical motion in the phase space. The orange dots corresponds to the times when the qubit is excited while the dark red dots corresponds to the times when the qubit is de-excited

thermal baths at infinite temperature to excite the qubit and at zero temperature to de-excite the qubit. In this latter case, the average work received by the MO over one period is $-\hbar\Delta\omega/2$. However both cases require a perfect synchronization of the excitation and de-excitation of the qubit with the mechanical motion. In this chapter we propose an autonomous alternative to this ideal energy converter.

5.1.2 Autonomous Converter

The situation under study, a non-isolated hybrid optomechanical system, is represented in Fig. 5.3a. A laser of frequency ω_L is shone on the qubit which is also in contact with an electromagnetic reservoir $\mathcal{R}_\mathrm{q}$ at zero temperature. We assume that the pure dephasing rate of the qubit is very large and chose the Rabi frequency g so that the qubit sees the laser as an incoherent source. Therefore, when interacting with the laser, the qubit ends in a mixed energy state and the laser can be assimilated to a hot bath with filtered frequency. Unlike the previous two chapters, we will study the evolution of the MO on time scales longer that the characteristic time of the mechanical damping Γ^{-1} and, therefore, take into account the reservoir $\mathcal{R}_\mathrm{m}$ of the MO.

The energy conversion principle is illustrated in Fig. 5.3b. The optomechanical frequency modulation makes the qubit periodically enter in resonance with the blue-detuned laser of frequency ω_L, at which point it can absorb a photon of energy $\hbar\omega_\mathrm{L}$. Then, by spontaneous emission, the qubit will emit a lower energy photon and the energy difference is given to the MO. In a similar way to three-level masers [1], the whole system can be seen as an autonomous thermal machine: the qubit receives heat from a hot source (the laser), gives part of this energy to the MO, which plays the role of the battery, and dumps the remaining energy in the cold bath (the electromagnetic

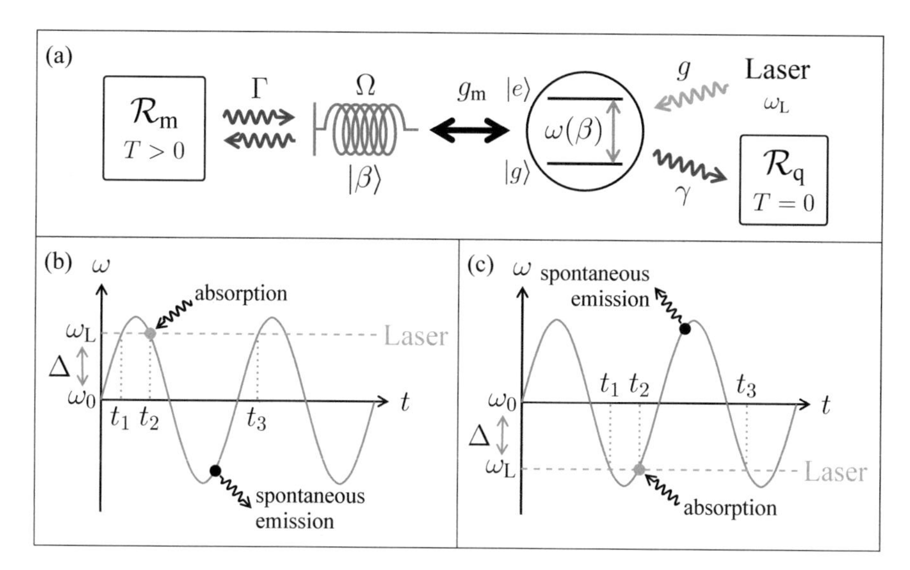

Fig. 5.3 Setup and principle of the energy conversion. **a** System under study: a qubit of frequency ω is parametrically coupled to a MO of frequency Ω with a coupling strength g_m. The MO is also coupled to a thermal bath $\mathcal{R}_m$ of finite temperature T with a damping rate Γ. The qubit interacts with a laser of frequency ω_L and an electromagnetic reservoir at zero temperature $\mathcal{R}_q$. g is the Rabi frequency and γ the spontaneous emission rate of the qubit. **b, c** Energy conversion principle for a blue-detuned laser and a red-detuned laser respectively. ω_0 is the transition frequency of the bare qubit. **b** Amplification principle: The qubit absorbs a high-energy photon from the laser and spontaneously emits a lower-energy one. The energy difference is transferred to the MO. **c** Cooling principle: the qubit absorbs a low-energy photon from the laser and emits a higher-energy one. The energy difference is provided by the MO

reservoir). The direction of the energy conversion is reversible: Using a red-detuned laser (Fig. 5.3c) results in the emission of higher energy photons by the qubit and the energy difference is provided by the MO which is therefore cooled down. As there is no need to connect/disconnect the qubit's baths, this energy converter is autonomous, unlike the ideal converter. The connection/disconnection of the hot bath happens by itself when the qubit's frequency moves in/out of resonance with the laser due to the optomechanical modulation.

Note that what we mean by cooling in the red detuning case is that the average phonon number in the MO decreases, but this not the usual thermodynamic definition of cooling. Indeed, as show in Fig. 5.1d, the heat is not flowing from the cold bath to the hot bath like in a classical refrigerator. Furthermore, we cannot associate a temperature to the MO, since it is not in a thermal state, as shown in Sect. 5.4.2.

5.2 Dynamics of the Mechanical Oscillator

To be able to understand more in detail the energy conversion process, we first study the dynamics of the MO. The evolution of the hybrid optomechanical system can be divided in two parts: when the emitter's frequency is too detuned to interact with laser and when the emitter is in resonance with the laser. The former part of the evolution, of characteristic time Ω^{-1}, is a lot longer than the latter which will be considered as instantaneous. We first derive a master equation for the hybrid optomechanical system and unravel it into quantum trajectories. Finally including the interaction between the qubit and the laser, we can study the stochastic evolution of the MO from these trajectories.

5.2.1 Evolution of the Hybrid Optomechanical System Without the Laser

In this section we do not take into account the interaction between the laser and the qubit. This corresponds to the first part of the evolution, the slowest one during which work is exchanged.

5.2.1.1 Microscopic Model

As in the previous chapters, the hybrid optomechanical system is described by the Hamiltonian (Eq. (3.1))

$$H_{\mathrm{qm}} = \hbar\omega_0 \left|e\right\rangle\!\left\langle e\right| \otimes \mathbf{1}_{\mathrm{m}}, + \mathbf{1}_{\mathrm{q}} \otimes \hbar\Omega b^\dagger b + \hbar g_{\mathrm{m}} \left|e\right\rangle\!\left\langle e\right| (b^\dagger + b). \tag{5.2}$$

The qubit is coupled to an electromagnetic reservoir $\mathcal{R}_{\mathrm{q}}$ at zero temperature characterized by the Hamiltonian (Eq. (3.13))

$$H_{\mathcal{R}_{\mathrm{q}}} = \sum_k \hbar\omega_k a_k^\dagger a_k. \tag{5.3}$$

The coupling between the qubit and $\mathcal{R}_{\mathrm{q}}$, denoted $V_{\mathcal{R}_{\mathrm{q}}} = \sum_{l=\pm} R_l \otimes \sigma_l$, with $R_+ = \sum_k \hbar g_k a_k$, $R_- = R_+^\dagger$. Reciprocally, the MO is coupled to a phonon reservoir $\mathcal{R}_{\mathrm{m}}$, but at finite temperature T, characterized by the Hamiltonian $H_{\mathcal{R}_{\mathrm{m}}}$ that reads

$$H_{\mathcal{R}_{\mathrm{m}}} = \sum_k \hbar\Omega_k c_k^\dagger c_k. \tag{5.4}$$

c_k is the annihilation operator of the k-th phononic mode of frequency Ω_k. $\mathcal{R}_{\mathrm{m}}$ is in a thermal state, therefore, it contains on average $\bar{N}_{\Omega_k}$ phonons at any frequency Ω_k, with

$$\bar{N}_{\Omega_k} = \left(\exp\left(\frac{\hbar \Omega_k}{k_B T} \right) - 1 \right)^{-1}. \tag{5.5}$$

The coupling Hamiltonian between the MO and $\mathcal{R}_m$, in the rotating wave approximation, equals

$$V_{\mathcal{R}_m} = \sum_{l=\pm} C_l \otimes b_l \tag{5.6}$$

where $C_+ = \sum_k \hbar G_k c_k$, $C_- = C_+^\dagger$ and G_k is the coupling strength between the MO and the k-th mode of the reservoir. For notation convenience, we have defined $b_- := b$ and $b_+ := b^\dagger$. We denote

$$\Gamma := \sum_k G_k^2 \delta(\Omega - \Omega_k) \tag{5.7}$$

the mechanical damping rate. Finally, the Hamiltonian for the total system writes

$$H_{\text{tot}} = H_{\text{qm}} + H_{\mathcal{R}_q} + H_{\mathcal{R}_m} + V_{\mathcal{R}_q} + V_{\mathcal{R}_m} \tag{5.8}$$

It can be split in two terms: $H_{\text{tot}} = H_0 + V$, with

$$H_0 := H_{\text{qm}} + H_{\mathcal{R}_q} + H_{\mathcal{R}_m}, \tag{5.9}$$

$$V := V_{\mathcal{R}_q} + V_{\mathcal{R}_m}. \tag{5.10}$$

In the interaction picture with respect to H_0, the density operator $\rho_{\text{tot}}^{\text{I}}$ of the total systems evolves according to

$$\dot{\rho}_{\text{tot}}^{\text{I}}(t) = -\frac{\text{i}}{\hbar}[V^{\text{I}}(t), \rho_{\text{tot}}^{\text{I}}(t)], \tag{5.11}$$

with $V^{\text{I}}(t) = V_{\mathcal{R}_q}^{\text{I}}(t) + V_{\mathcal{R}_m}^{\text{I}}(t)$. Defining $U_0(t) = \text{e}^{-\text{i}tH_0/\hbar}$, any operator A in the Schrödinger picture becomes $A^{\text{I}}(t) = U_0^\dagger(t) A U_0(t)$. In particular, the two coupling Hamiltonians become:

$$V_{\mathcal{R}_q}^{\text{I}}(t) = \sum_{l=\pm} R_l^{\text{I}}(t) \otimes \sigma_l^{\text{I}}(t), \tag{5.12}$$

$$V_{\mathcal{R}_m}^{\text{I}}(t) = \sum_{l=\pm} C_l^{\text{I}}(t) \otimes b_l^{\text{I}}(t), \tag{5.13}$$

with

$$R^{\mathrm{I}}_{-}(t) = \sum_k \hbar g_k a_k^{\dagger} \mathrm{e}^{\mathrm{i}\omega_k t}, \tag{5.14}$$

$$\sigma^{\mathrm{I}}_{-}(t) = \mathrm{e}^{-\mathrm{i}\omega_0 t}\sigma_{-} \exp(\mathrm{i}\Omega b^{\dagger} b t) \exp(-\mathrm{i}(\Omega b^{\dagger} b + g_{\mathrm{m}}(b + b^{\dagger}))t), \tag{5.15}$$

$$C^{\mathrm{I}}_{-}(t) = \sum_k \hbar G_k c_k^{\dagger} \mathrm{e}^{\mathrm{i}\Omega_k t}, \tag{5.16}$$

$$b^{\mathrm{I}}(t) = \mathrm{e}^{-\mathrm{i}\Omega t} b_{\mathrm{qm}} - \frac{g_{\mathrm{m}}}{\Omega} |e\rangle\langle e| . \tag{5.17}$$

We have defined

$$b_{\mathrm{qm}} := b\,|g\rangle\langle g| + \left(b + \frac{g_{\mathrm{m}}}{\Omega}\right) |e\rangle\langle e| . \tag{5.18}$$

5.2.1.2 Derivation of the Master Equation

We now derive the master equation describing the evolution of the optomechanical system. In the following, we consider the regime

$$\omega_0, (\tau_{\mathrm{c}}^{\mathrm{q}})^{-1}, (\tau_{\mathrm{c}}^{\mathrm{m}})^{-1} \gg g_{\mathrm{m}}, \Omega, \gamma \gg \Gamma, \tag{5.19}$$

where $\tau_{\mathrm{c}}^{\mathrm{q}}$ and $\tau_{\mathrm{c}}^{\mathrm{m}}$ are the correlation times of $\mathcal{R}_{\mathrm{q}}$ and $\mathcal{R}_{\mathrm{m}}$ respectively. This regime is fulfilled by the experimental implementations cited in Chap. 3 (See Table 3.1). We can therefore chose a time step Δt such that

$$\omega_0^{-1}, \tau_{\mathrm{c}}^{\mathrm{q}}, \tau_{\mathrm{c}}^{\mathrm{m}} \ll \Delta t \ll g_{\mathrm{m}}^{-1}, \Omega^{-1}, \gamma^{-1}, \Gamma^{-1}. \tag{5.20}$$

Applying the Born-Markov approximation, like in Sect. 3.1.3, the precursor of the master equation, derived from Eq. (5.11), reads

$$\begin{aligned}
\Delta\rho^{\mathrm{I}}_{\mathrm{qm}}(t) &= \mathrm{Tr}_{\mathcal{R}_{\mathrm{q}},\mathcal{R}_{\mathrm{m}}}(\Delta\rho^{\mathrm{I}}_{\mathrm{tot}}(t)) \\
&= -\frac{\mathrm{i}}{\hbar}\int_t^{t+\Delta t} \mathrm{d}t' \mathrm{Tr}_{\mathcal{R}_{\mathrm{q}},\mathcal{R}_{\mathrm{m}}}\left([V^{\mathrm{I}}(t'), \rho^{\mathrm{I}}_{\mathrm{qm}}(t) \otimes \rho_{\mathcal{R}_{\mathrm{q}}} \otimes \rho_{\mathcal{R}_{\mathrm{m}}}]\right) \\
&\quad - \frac{1}{\hbar^2}\int_t^{t+\Delta t} \mathrm{d}t' \int_t^{t'} \mathrm{d}t'' \mathrm{Tr}_{\mathcal{R}_{\mathrm{q}},\mathcal{R}_{\mathrm{m}}}\left(\left[V^{\mathrm{I}}(t'), \left[V^{\mathrm{I}}(t''), \rho^{\mathrm{I}}_{\mathrm{qm}}(t) \otimes \rho_{\mathcal{R}_{\mathrm{q}}} \otimes \rho_{\mathcal{R}_{\mathrm{m}}}\right]\right]\right),
\end{aligned} \tag{5.21}$$

As the reservoirs are in thermal states, $\mathrm{Tr}_{\mathcal{R}_{\mathrm{q}}}(R^{\mathrm{I}}_{\pm}(t)\rho_{\mathcal{R}_{\mathrm{q}}}) = 0$ and $\mathrm{Tr}_{\mathcal{R}_{\mathrm{m}}}(C^{\mathrm{I}}_{\pm}(t)\rho_{\mathcal{R}_{\mathrm{m}}}) = 0$. Therefore, all terms containing a single $V^{\mathrm{I}}_{\mathcal{R}_{\mathrm{q}}}(t)$ or $V^{\mathrm{I}}_{\mathcal{R}_{\mathrm{m}}}(t)$ vanish, giving

$$\Delta\rho^{\mathrm{I}}_{\mathrm{qm}}(t) = -\frac{1}{\hbar^2}\int_t^{t+\Delta t}\mathrm{d}t'\int_t^{t'}\mathrm{d}t''\Big(\mathrm{Tr}_{\mathcal{R}_{\mathrm{q}}}\Big(\Big[V^{\mathrm{I}}_{\mathcal{R}_{\mathrm{q}}}(t'),\Big[V^{\mathrm{I}}_{\mathcal{R}_{\mathrm{q}}}(t''),\rho^{\mathrm{I}}_{\mathrm{qm}}(t)\otimes\rho_{\mathcal{R}_{\mathrm{q}}}\Big]\Big]\Big) + \mathrm{Tr}_{\mathcal{R}_{\mathrm{m}}}\Big(\Big[V^{\mathrm{I}}_{\mathcal{R}_{\mathrm{m}}}(t'),\Big[V^{\mathrm{I}}_{\mathcal{R}_{\mathrm{m}}}(t''),\rho^{\mathrm{I}}_{\mathrm{qm}}(t)\otimes\rho_{\mathcal{R}_{\mathrm{m}}}\Big]\Big]\Big)\Big). \tag{5.22}$$

Expanding the commutators, the trace over the Hilbert space of $\mathcal{R}_{\mathrm{q}}$ gives correlation functions $g_{ll'}(u,v)$, defined by Eq. (3.33), where u, v are two times and $l, l' \in \{+,-\}$. This reservoir is at zero temperature so all these functions equal zero except $g_{--}(u,v)$ which reads

$$g_{--}(u,v) = \hbar^2\sum_k g_k^2 \mathrm{e}^{-\mathrm{i}\omega_k(u-v)}. \tag{5.23}$$

Similarly, the trace over the Hilbert space of $\mathcal{R}_{\mathrm{m}}$ yields terms of the form

$$G_{ll'}(u,v) := \mathrm{Tr}_{\mathcal{R}_{\mathrm{m}}}(\rho_{\mathcal{R}_{\mathrm{m}}} C^{\mathrm{I}}_l(u)^\dagger C^{\mathrm{I}}_{l'}(v)). \tag{5.24}$$

If $l \neq l'$ this trace vanishes, otherwise, we get the two correlation functions:

$$G_{--}(u,v) = \hbar^2\sum_k g_k^2(\bar{N}_{\Omega_k}+1)\mathrm{e}^{-\mathrm{i}\Omega_k(u-v)}, \tag{5.25}$$

$$G_{++}(u,v) = \hbar^2\sum_k g_k^2\bar{N}_{\Omega_k}\mathrm{e}^{\mathrm{i}\Omega_k(u-v)}. \tag{5.26}$$

The integral $\int_t^{t'}\mathrm{d}t''$ can then be changed into an integral over $\tau = t'-t''$: $\int_0^{t'-t}\mathrm{d}\tau$. As $g_{--}(u,v) = g_{--}(u-v)$ is non zero only for $|u-v| \lesssim \tau_c^{\mathrm{q}} \ll \Delta t$, and $G_{ll}(u,v) = G_{ll}(u-v)$ for $|u-v| \lesssim \tau_c^{\mathrm{m}} \ll \Delta t$, the upper bound can be set to infinity [8]:

$$\begin{aligned}\Delta\rho^{\mathrm{I}}_{\mathrm{qm}}(t) = -\frac{1}{\hbar^2}\int_t^{t+\Delta t}\mathrm{d}t'\int_0^{\infty}\mathrm{d}\tau\Big[&g_{--}(\tau)\big(\sigma^{\mathrm{I}}_+(t')\sigma^{\mathrm{I}}_-(t'-\tau)\rho^{\mathrm{I}}_{\mathrm{qm}}(t) \\ &- \sigma^{\mathrm{I}}_-(t'-\tau)\rho^{\mathrm{I}}_{\mathrm{qm}}(t)\sigma^{\mathrm{I}}_+(t')\big) \\ &+\sum_{l=\pm}G_{ll}(\tau)\big(b^{\mathrm{I}}_l(t')^\dagger b^{\mathrm{I}}_l(t'-\tau)\rho^{\mathrm{I}}_{\mathrm{qm}}(t) \\ &- b^{\mathrm{I}}_l(t'-\tau)\rho^{\mathrm{I}}_{\mathrm{qm}}(t)b^{\mathrm{I}}_l(t')^\dagger\big)\Big] \\ &+\mathrm{h.c.}\end{aligned} \tag{5.27}$$

Using Eqs. (3.37), that is

$$\sigma^{\mathrm{I}}_+(t'-\tau) = \sigma^{\mathrm{I}}_+(t')\mathrm{e}^{-\mathrm{i}\tilde{\omega}\tau}, \tag{5.28}$$

$$\sigma^{\mathrm{I}}_-(t'-\tau) = \mathrm{e}^{\mathrm{i}\tilde{\omega}\tau}\sigma^{\mathrm{I}}_-(t'), \tag{5.29}$$

with

$$\tilde{\omega} := \omega_0 \mathbf{1}_\mathrm{m} + g_\mathrm{m} \left(b\mathrm{e}^{-\mathrm{i}\Omega t} + b^\dagger \mathrm{e}^{\mathrm{i}\Omega t} \right), \tag{5.30}$$

and Eq. (5.17), we obtain

$$\begin{aligned}\Delta\rho_\mathrm{qm}^\mathrm{I}(t) = -\frac{1}{\hbar^2}\int_t^{t+\Delta t}\mathrm{d}t'\int_0^\infty \mathrm{d}\tau\Big[& g_{--}(\tau)\big(\sigma_+^\mathrm{I}(t')\mathrm{e}^{\mathrm{i}\tilde{\omega}\tau}\sigma_-^\mathrm{I}(t')\rho_\mathrm{qm}^\mathrm{I}(t) - \sigma_-^\mathrm{I}(t')\rho_\mathrm{qm}^\mathrm{I}(t)\sigma_+^\mathrm{I}(t')\big) \\ & + \sum_{l=\pm} G_{ll}(\tau)\mathrm{e}^{-\mathrm{i}l\Omega\tau}\left((b_\mathrm{qm}^l)^\dagger b_\mathrm{qm}^l\rho_\mathrm{qm}^\mathrm{I}(t) - b_\mathrm{qm}^l\rho_\mathrm{qm}^\mathrm{I}(t)(b_\mathrm{qm}^l)^\dagger\right) \\ & + \sum_{l=\pm} G_{ll}(\tau)\mathrm{e}^{\mathrm{i}l\Omega(t'-\tau)}\frac{g_\mathrm{m}}{\Omega}\left(-|e\rangle\langle e|\, b_\mathrm{qm}^l\rho_\mathrm{qm}^\mathrm{I}(t) + b_\mathrm{qm}^l\rho_\mathrm{qm}^\mathrm{I}(t)\,|e\rangle\langle e|\right)\Big] \\ + \mathrm{h.c.} & \end{aligned} \tag{5.31}$$

We have defined $b_\mathrm{qm}^- := b_\mathrm{qm}$, $b_\mathrm{qm}^+ := b_\mathrm{qm}^\dagger$ for notation convenience. The extra terms coming from Eq. (5.17) do not depend on τ, except for the $G_{ll}(\tau)$ factor, so the integral over $G_{ll}(\tau)$ gives zero since there are no phonons at zero frequency. As $\Delta t \ll g_\mathrm{m}^{-1}, \Omega^{-1}$, integrating over t' approximately gives

$$\begin{aligned}\dot{\rho}_\mathrm{qm}^\mathrm{I}(t) &= \frac{\Delta\rho_\mathrm{qm}^\mathrm{I}}{\Delta t}(t) \\ &= -\frac{1}{\hbar^2}\int_0^\infty \mathrm{d}\tau\Big[g_{--}(\tau)\big(\sigma_+^\mathrm{I}(t)\mathrm{e}^{\mathrm{i}\tilde{\omega}\tau}\sigma_-^\mathrm{I}(t)\rho_\mathrm{qm}^\mathrm{I}(t) - \sigma_-^\mathrm{I}(t)\rho_\mathrm{qm}^\mathrm{I}(t)\sigma_+^\mathrm{I}(t)\big) \\ &\quad + \sum_{l=\pm} G_{ll}(\tau)\mathrm{e}^{-\mathrm{i}l\Omega\tau}\left((b_\mathrm{qm}^l)^\dagger b_\mathrm{qm}^l\rho_\mathrm{qm}^\mathrm{I}(t) - b_\mathrm{qm}^l\rho_\mathrm{qm}^\mathrm{I}(t)(b_\mathrm{qm}^l)^\dagger\right) \\ &\quad + \sum_{l=\pm} G_{ll}(\tau)\mathrm{e}^{\mathrm{i}l\Omega(t-\tau)}\frac{g_\mathrm{m}}{\Omega}\left(-|e\rangle\langle e|\, b_\mathrm{qm}^l\rho_\mathrm{qm}^\mathrm{I}(t) + b_\mathrm{qm}^l\rho_\mathrm{qm}^\mathrm{I}(t)\,|e\rangle\langle e|\right)\Big] \\ &\quad + \mathrm{h.c.}\end{aligned} \tag{5.32}$$

Assuming that the frequency variation of the qubit due to the optomechanical coupling does not change the spontaneous emission rate like in Chap. 3 (Eq. (3.49)), the first term gives rise to the usual Lindbladian $\mathcal{L}_\mathrm{q}$ for a qubit in contact with a zero temperature bath and does not affect the mechanical state,

$$\mathcal{L}_\mathrm{q}[\rho_\mathrm{qm}(t)] = \gamma D[\sigma_- \otimes \mathbf{1}_\mathrm{m}]\rho_\mathrm{qm}(t). \tag{5.33}$$

The second term yields a Lindbladian $\mathcal{L}_\mathrm{m}$ similar to the one of a harmonic oscillator in contact with a thermal bath, except that the annihilation operator is replaced by b_qm,

$$\mathcal{L}_\mathrm{m}[\rho_\mathrm{qm}(t)] = \Gamma N_\mathrm{th} D[b_\mathrm{qm}^\dagger]\rho_\mathrm{qm}(t) + \Gamma(N_\mathrm{th}+1)D[b_\mathrm{qm}]\rho_\mathrm{qm}(t), \tag{5.34}$$

with $N_\mathrm{th} := \bar{N}_\Omega$. Because of the last term, Eq. (5.31) cannot be put in the form of a Lindblad master equation. However, this last term only contributes to the evolution of the off-diagonal terms of the qubit's density operator in the $\{|e\rangle, |g\rangle\}$ basis and, due to the form of the Hamiltonian H_{qm} (Eq. (3.6)) and of the couplings to the bath, in the absence of coherence in the initial state, none will be built. Since in the following we will always start with the qubit in a mixed energy state containing no coherence and since the excitation by the laser in performed incoherently, we can safely neglect this contribution. Finally the evolution of the optomechanical system, in the Schrödinger picture, can be written in the form of the Lindblad master equation

$$\dot{\rho}_{qm}(t) = -\frac{i}{\hbar}[H_{qm}, \rho_{qm}(t)] + \mathcal{L}_q[\rho_{qm}(t)] + \mathcal{L}_m[\rho_{qm}(t)]. \tag{5.35}$$

5.2.1.3 Quantum Trajectories

The master equation (5.35) can be unraveled using quantum jumps (See Sect. 2.1.2.1) for $\mathcal{L}_q$, which amounts to detect the photons emitted by the qubit, and quantum state diffusion for $\mathcal{L}_m$ (See Sect. 2.1.2.2). Indeed, the Kraus operator corresponding to spontaneous emission is $M_{sp} = \sqrt{\gamma \Delta t}\sigma_-$ and the no-jump part is $M_0 = 1 - i\Delta t H_{qm}/\hbar - \gamma\Delta t/2\,|e\rangle\langle e|$. Since the quantum state diffusion Kraus operators M_r, such that

$$\int \mathrm{d}r\, M_r \rho_{qm} M_r^\dagger = \mathcal{L}_m[\rho_{qm}]\Delta t \tag{5.36}$$

obeys the normalization condition Eq. (2.3), the no-jump part can be further decomposed as

$$M_0 = \int \mathrm{d}r\, M_r \left(1 - \frac{i\Delta t}{\hbar} H_{qm} - \frac{\gamma \Delta t}{2}\,|e\rangle\langle e|\right). \tag{5.37}$$

Therefore the set of Kraus operators composed of M_{sp} and $\{M_r(1 - i\Delta t H_{qm}/\hbar - \gamma\Delta t/2\,|e\rangle\langle e|)\}_r$ is a well defined Kraus decomposition and can be used to unravel the master equation.

Denoting $\vec{\Sigma}$ a trajectory and $|\Psi_\Sigma(t)\rangle$ the state of the hybrid system at time t, either the qubit emits a photon during the time step $[t, t+\mathrm{d}t[$ and

$$|\Psi_\Sigma(t+\mathrm{d}t)\rangle = \frac{1}{|\langle e|\,\Psi_\Sigma(t)\rangle|}\,|g\rangle \otimes \langle e|\,\Psi_\Sigma(t)\rangle\,, \tag{5.38}$$

with probability $\gamma \mathrm{d}t|\langle e|\,\Psi_\Sigma(t)\rangle|^2$ or, with probability $1 - \gamma\mathrm{d}t|\langle e|\,\Psi_\Sigma(t)\rangle|^2$, the system evolves according to the quantum state diffusion equation (in Itô form [9]):

$$
\begin{aligned}
\mathrm{d}\left|\Psi_{\Sigma}(t)\right\rangle = \Bigg\{ & -\frac{\mathrm{i}}{\hbar} H_{\mathrm{qm}} \mathrm{d}t - \frac{\Gamma N_{\mathrm{th}}}{2}\left(b_{\mathrm{qm}} b_{\mathrm{qm}}^{\dagger} + \left|\left\langle b_{\mathrm{qm}}\right\rangle_{\Psi_{\Sigma}(t)}\right|^{2} - 2\left\langle b_{\mathrm{qm}}\right\rangle_{\Psi_{\Sigma}(t)} b_{\mathrm{qm}}^{\dagger}\right) \mathrm{d}t \\
& -\frac{\Gamma(N_{\mathrm{th}}+1)}{2}\left(b_{\mathrm{qm}}^{\dagger} b_{\mathrm{qm}} + \left|\left\langle b_{\mathrm{qm}}\right\rangle_{\Psi_{\Sigma}(t)}\right|^{2} - 2\left\langle b_{\mathrm{qm}}^{\dagger}\right\rangle_{\Psi_{\Sigma}(t)} b_{\mathrm{qm}}\right) \mathrm{d}t \\
& +\sqrt{\Gamma N_{\mathrm{th}}}\left(b_{\mathrm{qm}}^{\dagger} - \left\langle b_{\mathrm{qm}}^{\dagger}\right\rangle_{\Psi_{\Sigma}(t)}\right) \mathrm{d}\xi_{+}(t) \\
& +\sqrt{\Gamma(N_{\mathrm{th}}+1)}\left(b_{\mathrm{qm}} - \left\langle b_{\mathrm{qm}}\right\rangle_{\Psi_{\Sigma}(t)}\right) \mathrm{d}\xi_{-}(t) \Bigg\} \left|\Psi_{\Sigma}(t)\right\rangle
\end{aligned} \tag{5.39}
$$

where $\langle . \rangle_{\Psi_{\Sigma}(t)} = \langle . \rangle \, \Psi_{\Sigma}(t)$ is the expectation value of the operator at time t, $\mathrm{d}\xi_{-}$ and $\mathrm{d}\xi_{+}$ are two independent complex Wiener increments.

We now show that the optomechanical system is always in a factorized state along a given trajectory $\vec{\Sigma}$, consisting of an energy eigenstate of the qubit and a coherent mechanical state. Let's assume that at time t, the system is in state $|\Psi_{\Sigma}(t)\rangle = |\epsilon, \beta\rangle$ where $\epsilon \in \{e, g\}$ and $|\beta\rangle$ is a coherent state of the MO. Then during the next time step $\mathrm{d}t$, either the qubit emits a photon and $|\Psi_{\Sigma}(t+\mathrm{d}t)\rangle = |g, \beta\rangle$ with probability $\gamma \mathrm{d}t \delta_{\epsilon,e}$, or with probability $1 - \gamma \mathrm{d}t \delta_{\epsilon,e}$, the system evolves according to Eq. (5.39) which can be rewritten at the first order in $\mathrm{d}t$:

$$
\begin{aligned}
|\Psi_{\Sigma}(t+\mathrm{d}t)\rangle &= \exp(-\mathrm{i}\left(\omega_0 - \frac{g_{\mathrm{m}}^2}{\Omega}\right)\mathrm{d}t\delta_{\epsilon,e} - \frac{\Gamma N_{\mathrm{th}}}{2}\mathrm{d}t + \frac{\Gamma}{2}|\beta_\epsilon|^2 \mathrm{d}t) \\
&\quad \times \exp\left(\sqrt{\Gamma N_{\mathrm{th}}}(b_{\mathrm{qm}}^{\dagger} - \beta_\epsilon^*)\mathrm{d}\xi_{+}(t) - \left(\mathrm{i}\Omega + \frac{\Gamma}{2}\right)\mathrm{d}t b_{\mathrm{qm}}^{\dagger} b_{\mathrm{qm}}\right)|\epsilon, \beta\rangle \\
&= \exp\left(-\mathrm{i}\left(\omega_0 - \frac{g_{\mathrm{m}}^2}{\Omega}\right)\mathrm{d}t\delta_{\epsilon,e} - \frac{\Gamma N_{\mathrm{th}}}{2}\mathrm{d}t\right) \\
&\quad \times \exp\left(\sqrt{\Gamma N_{\mathrm{th}}}(b_{\mathrm{qm}}^{\dagger} - \beta_\epsilon^*)\mathrm{d}\xi_{+}(t)\right)\left|\epsilon, \beta_\epsilon \mathrm{e}^{-(\mathrm{i}\Omega + \frac{\Gamma}{2})\mathrm{d}t} - \frac{g_{\mathrm{m}}}{\Omega}\delta_{\epsilon,e}\right\rangle
\end{aligned} \tag{5.40}
$$

where $\beta_\epsilon = \beta + \frac{g_{\mathrm{m}}}{\Omega}\delta_{\epsilon,e}$ is the eigenvalue of b_{qm} associated with the eigenstate $|\epsilon, \beta\rangle$. From this expression, it can be checked that

$$
b\,|\Psi_{\Sigma}(t+\mathrm{d}t)\rangle = \left(\beta_\epsilon \mathrm{e}^{-(\mathrm{i}\Omega + \frac{\Gamma}{2})\mathrm{d}t} - \frac{g_{\mathrm{m}}}{\Omega}\delta_{\epsilon,e} + \sqrt{\Gamma N_{\mathrm{th}}}\mathrm{d}\xi_{+}(t)\right)|\Psi(t+\mathrm{d}t)\rangle\,. \tag{5.41}
$$

Therefore, the MO remains in a coherent state while the qubit's state is unchanged. As a consequence, if at time $t = 0$ the hybrid system is prepared in state $|\epsilon_{\Sigma}(0), \beta_{\Sigma}(0)\rangle$, then at any time t, its state is still of the same form and can be denoted $|\epsilon_{\Sigma}(t), \beta_{\Sigma}(t)\rangle$. In the absence of spontaneous emission, the state of the MO at time $t + \mathrm{d}t$ is related to the one at time t by the equation

$$
\beta_{\Sigma}(t+\mathrm{d}t) = \left(\beta_{\Sigma}(t) + \frac{g_{\mathrm{m}}}{\Omega}\delta_{\epsilon_{\Sigma}(t),e}\right)\mathrm{e}^{-(\mathrm{i}\Omega + \frac{\Gamma}{2})\mathrm{d}t} - \frac{g_{\mathrm{m}}}{\Omega}\delta_{\epsilon_{\Sigma}(t),e} + \sqrt{\Gamma N_{\mathrm{th}}}\mathrm{d}\xi_{+}(t). \tag{5.42}
$$

Therefore, at all times, the MO remains in a coherent state and the whole hybrid optomechanical system is in a pure product state, denoted $|\epsilon_\Sigma(t), \beta_\Sigma(t)\rangle$. This generalizes the proof from Chap. 4 (Sect. 4.1.1) to the case where the environment of the MO is taken into account in the form of quantum state diffusion.

5.2.2 *Interaction Between the Qubit and the Laser*

So far, the qubit was only in contact with a bath at zero temperature, thus at most one quantum jump could happen per trajectory. We now take into account the interaction between the laser and qubit. We assume that we are in the saturated regime of the Rabi oscillations, so that the laser can be considered as an incoherent source. Therefore, after its interaction with the laser, the qubit is prepared in a mixed state

$$\rho_q(\theta) = \theta \,|e\rangle\langle e| + (1-\theta)\,|g\rangle\langle g| \,, \tag{5.43}$$

where $\theta \in [0, 1]$ is called the excitation rate in the following. In the quantum trajectory framework, this translates into the qubit being in the excited state (resp. ground state) with probability θ (resp. $1-\theta$) after each interaction with the laser.

To summarize, when $\omega(\beta_\Sigma(t)) = \omega_{\mathrm{L}}$, the qubit enters in resonance with the laser and can absorb a photon with probability θ. In between resonances, apart from spontaneous emission, the qubit state remains the same while the MO evolves according to Eq. (5.42). This evolution can be numerically simulated. As it involves many different parameters, we first identify the interesting regimes by studying a simple coarse-grained model in the next section.

5.3 Coarse-Grained Model

This model describes the coarse-grained evolution of the average phonon number $N(t)$ in the MO. It enables us to determine the phonon number in the steady state by a fixed point study, as explained below.

5.3.1 *Evolution of the Phonon Number*

We derive this model by studying the variations of N over one mechanical period. The MO interacts with a thermal bath and the qubit, therefore, the variations of $N(t)$ can be split in two contributions:

$$\dot{N} = -f_{\mathrm{th}}(N) + \alpha_{\mathrm{m}}(N). \tag{5.44}$$

The thermal contribution $f_{\mathrm{th}}(N)$ is defined by

$$f_{\mathrm{th}}(N) := \Gamma(N - N_{\mathrm{th}}), \tag{5.45}$$

and $\alpha_{\mathrm{m}}(N)$ denotes the optomechanical contribution. $-\Gamma N$ corresponds to the losses caused by the mechanical damping and ΓN_{th} is the thermal source term, namely the phonons coming from the bath.

By analogy with lasers [10], we define the gain per mechanical period

$$G_{\mathrm{m}} := \frac{1}{N}\frac{\Delta N_{\mathrm{m}}}{T_{\mathrm{m}}}, \tag{5.46}$$

where $T_{\mathrm{m}} = 2\pi/\Omega$ is the mechanical period and ΔN_{m} is the variation of the phonon number during one period in the absence of thermal bath. Therefore, the optomechanical contribution reads

$$\alpha_{\mathrm{m}}(N) = G_{\mathrm{m}}(N)N. \tag{5.47}$$

The mechanical energy variation over one period is approximately given by $\Delta\varepsilon_{\mathrm{m}} \simeq \hbar\Omega\Delta N_{\mathrm{m}}$ because $\Gamma \ll \Omega$. As the energy balance gives $\Delta\varepsilon_{\mathrm{m}} = W$ (Eq. (3.73)), where W the average work received by the MO, the gain can be expressed as

$$G_{\mathrm{m}} = \frac{1}{N}\frac{\Delta\varepsilon_{\mathrm{m}}}{\hbar\Omega T_{\mathrm{m}}} = \frac{1}{N}\frac{W}{\hbar 2\pi}. \tag{5.48}$$

Therefore, $\alpha_{\mathrm{m}}(N)$ can be obtained by calculating the amount of work exchanged between the MO and the qubit over one mechanical period. Over this time scale, the qubit transition frequency can be approximated by

$$\omega(t) = \omega_0 + 2g_{\mathrm{m}}\sqrt{N(t_0)}\sin(\Omega(t - t_0)), \tag{5.49}$$

where t_0 is the start of the considered mechanical period and $t \in [t_0, t_0 + 2\pi/\Omega]$. We will now compute the average work, using that $W = \left\langle W[\vec{\Sigma}]\right\rangle_{\vec{\Sigma}}$, where $W[\vec{\Sigma}]$ is the work received by the MO over one period for one given trajectory $\vec{\Sigma}$ of the optomechanical system.

As in Chap. 4, at the single trajectory level, the energy of the hybrid system $\varepsilon_{\mathrm{qm}}[\vec{\Sigma}, t]$, given by Eq. (4.18), naturally splits into two distinct components respectively quantifying the qubit and the mechanical energies: $\varepsilon_{\mathrm{q}}[\vec{\Sigma}, t]$ (Eq. (4.19)) and $\varepsilon_{\mathrm{m}}[\vec{\Sigma}, t]$ (Eq. (4.20)). Since we are modeling the interaction between the qubit and the laser in an effective way, the variation of ε_{q} due to this interaction is identified with heat exchanges with the hot bath. Similarly, the variation of ε_{q} due to spontaneous emission corresponds to heat exchanges with the cold bath. In between such events, the state of the qubit does not change, only its effective frequency changes, therefore the variation of ε_{q} corresponds to work, given by Eq. (2.42). Integrating the work increment, $W[\vec{\Sigma}]$ reads

$$W[\vec{\Sigma}] = -\int_{t_1}^{t_3} \mathrm{d}t \delta_{\epsilon_\Sigma(t),e} \hbar\dot{\omega}(t), \tag{5.50}$$

where we have defined t_i, $i = 1, 2, 3$, as the times at which the qubit enters in resonance with the laser (See Fig. 5.3b and c):

$$t_1 = \frac{1}{\Omega} \arcsin\left(\frac{\Delta}{2g_\mathrm{m}\sqrt{N}}\right), \tag{5.51}$$

$$t_2 = \frac{\pi}{\Omega} - t_1, \tag{5.52}$$

$$t_3 = t_1 + \frac{2\pi}{\Omega}. \tag{5.53}$$

We have denoted

$$\Delta := \omega_\mathrm{L} - \omega_0 \tag{5.54}$$

the detuning between the laser and the frequency of the bare qubit. The work can be split in two parts: $W_1[\vec{\Sigma}]$ and $W_2[\vec{\Sigma}]$, where $W_i[\vec{\Sigma}]$, with $i = 1, 2$, is the work received by the MO between t_i and t_{i+1}. To obtain the average work $W = W_1 + W_2$, we average on the state of the qubit after its interaction with the laser and, if the qubit is in the excited state, on the spontaneous emission time t_sp, therefore

$$W_i = \theta \int_{t_i}^{t_{i+1}} \mathrm{d}t_\mathrm{sp}\, \gamma \mathrm{e}^{-\gamma(t_\mathrm{sp}-t_i)} \hbar(\omega(t_i) - \omega(t_\mathrm{sp})). \tag{5.55}$$

Finally, we obtain

$$\alpha_\mathrm{m}(N) = \Big(\Omega\Delta\left(1 - \mathrm{e}^{-\gamma\pi/\Omega}\cosh(2\gamma t_1)\right) - \gamma 2g_\mathrm{m}\sqrt{N - N_\mathrm{min}}\,\mathrm{e}^{-\gamma\pi/\Omega}\sinh(2\gamma t_1)\Big) \times \frac{\theta\Omega}{\pi(\gamma^2 + \Omega^2)}\Theta(N - N_\mathrm{min}), \tag{5.56}$$

where Θ is the Heaviside function. We have defined the phonon number

$$N_\mathrm{min} := \left(\frac{\Delta}{2g_\mathrm{m}}\right)^2, \tag{5.57}$$

which is a threshold of $\alpha_\mathrm{m}(N)$. Indeed, when $N < N_\mathrm{min}$, the qubit is never in resonance with the laser (See Fig. 5.3b and c) so there is no optomechanical energy conversion and $\alpha_\mathrm{m}(N) = 0$. Conversely, if $N > N_\mathrm{min}$, $\alpha_\mathrm{m}(N)$ is non zero because of the work exchanges between the qubit and the MO.

5.3.2 Discussion

Using Eqs. (5.44) and (5.56), we can find the fixed points of N and determine the interesting values of the parameters for both the amplification and the cooling. Figure 5.4 presents the results of this fixed point study. The three possible situations for blue detuning, $\Delta > 0$, are represented in Fig. 5.4a, b and c. If the gain term α_m is too weak or N_{min} too large, then the damping dominates and there is a single stable fixed point corresponding to the thermal phonon number N_{th} (Fig. 5.4a). Otherwise, a stable fixed point with a larger phonon number N_{st} exists (Fig. 5.4b, c) and the mechanical amplification is possible. However, the (c) case does not give access to large phonon numbers for the fixed point so in the following we will target the case (b). This latter case exhibits two stable fixed points on both side of a flow separation line in N_{min}. The energy conversion process can be reversed by using a red detuning, $\Delta < 0$. The graphical study in Fig. 5.4d, e and f shows that there is at most one fixed point whose position mostly depends on how N_{min} compares to the thermal phonon number N_{th}. More precisely, we need to have $N_{min} < N_{th}$ to be able to cool down the MO.

We also studied the impact of γ on the energy conversion. We determined that outside the regime $\gamma \sim \Omega$, the energy conversion does not work. This can be understood from Fig. 5.3b and c: If the spontaneous emission rate is too large, the emission takes place right after the absorption and barely any work is exchanged with the MO, so the conversion efficiency is too low to overcome thermal noise. Conversely, if the spontaneous emission rate is too small, the emission often does not occur before the qubit interacts with the laser again, therefore the total work is zero. Amplification of the mechanical motion in the regime $\gamma \gg \Omega$ was proposed in Ref. [11], but it requires a modulation of the optical drive at the mechanical frequency. On the contrary, having $\gamma \sim \Omega$ gives rise to an autonomous modulation: The qubit enters in and out of resonance with the laser due to the optomechanical coupling and the spontaneous emission occurs most of the time near the targeted extremal frequency. More precisely, for a blue detuning, in the case corresponding to Fig. 5.4b, the value of γ maximizing N_{st} is $\gamma \simeq \Omega/2$, regardless of the other parameters. The value of γ has less influence on N_{st} for red detunings, but $\gamma \simeq \Omega/2$ is also in the range of values that give the lowest phonon numbers. We will therefore chose $\gamma = \Omega/2$ in the following. The spontaneous emission rate of the qubit in the devices mentioned in Table 3.1 are not close to this value. However in Ref. [12], γ is smaller than Ω so it should be possible to make the lifetime of the transmon qubit shorter. In Ref. [13] it should also be possible to have $\gamma \sim \Omega$ by replacing the InAs quantum dot by one with a longer lifetime, such as the ones in Refs. [14, 15].

Finally, having a large optomechanical coupling ratio g_m/Ω is less crucial than in Chap. 4, $g_m \sim \Omega$ is sufficient to get a noticeable change in the phonon number in the MO. Nevertheless, for blue detunings, N_{st} increases with g_m. After identifying the interesting regimes for both the cooling and the amplification, in the next section, we use numerically generated trajectories of the optomechanical system to characterize more precisely the behavior of the MO.

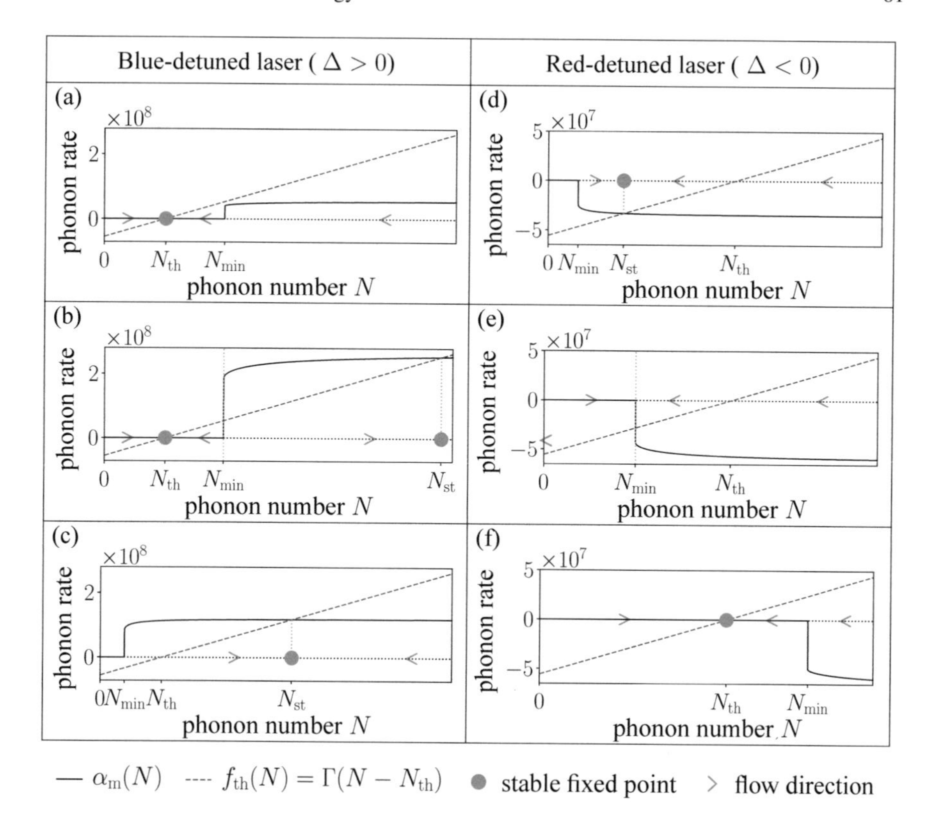

Fig. 5.4 Fixed point study: The thermal contribution $f_{\mathrm{th}}(N)$ (dotted red line) and the optomechanical one $\alpha_{\mathrm{m}}(N)$ (solid green line) are plotted in order to determine the sign of $\dot{N}$, hence the direction of the flow (gray arrows). *Left column:* Blue detuning. **a** $N_{\mathrm{th}} < N_{\mathrm{min}}$ and α_{m} is always below f_{th}: a single fixed point at the thermal phonon number. **b** $N_{\mathrm{th}} < N_{\mathrm{min}}$ and α_{m} intersects f_{th}: two stable fixed points, one at N_{th} and the other above, at N_{st}, and a flow separation line at $N = N_{\mathrm{min}}$. **c** $N_{\mathrm{th}} > N_{\mathrm{min}}$, so α_{m} intersects f_{th}: a single stable fixed point N_{st} above the thermal phonon number. *Right column:* Red detuning. **d** $N_{\mathrm{th}} > N_{\mathrm{min}}$ and α_{m} intersects f_{th}: a single fixed point N_{st} below the thermal phonon number. **e** $N_{\mathrm{th}} > N_{\mathrm{min}}$ and α_{m} does not intersect f_{th} due to the discontinuity in N_{min}: no fixed point, but the flows go towards N_{min}. **f** $N_{\mathrm{th}} < N_{\mathrm{min}}$: a single fixed point at the thermal phonon number

5.4 Characterization of the Energy Conversion

We first explore the blue detuning case, evidencing a laser-like behavior. Then we consider the red detuning case and investigate the mechanism for reducing the mechanical energy. Finally, we study the energy conversion efficiency.

5.4.1 Blue-Detuned Laser: Laser-Like Behavior

In this part, we consider a blue detuning, $\Delta > 0$. To investigate further the amplification predicted by the fixed point study (Fig. 5.4, left column), we numerically generated quantum trajectories for different excitation rates θ. Examples of obtained trajectories $\beta_\Sigma(t)$ are given in Fig. 5.5a, b and c. For the smallest value of θ (blue curve in Fig. 5.5b), the amplification is dominated by the thermal noise and nothing happens, as predicted by the coarse-grained model (Fig. 5.4a). For larger values of θ (corresponding to Fig. 5.4b), the amplification is visible. However, for $\theta = 0.1$, the phonon number fluctuates in between the two fixed points (orange curve in Fig. 5.5b). This is because the distance between the fixed point and the flow separation line is small enough to be crossed by thermal fluctuations. Conversely, if the detuning is very large, the thermal fluctuations will never cross the flow line and the MO will remain at the thermal fixed point.

The phase $\phi_\Sigma(t)$ of $\beta_\Sigma(t)$, after taking out the part rotating at the effective mechanical frequency Ω_{eff}, is represented in Fig. 5.5c. The effective mechanical frequency Ω_{eff} is not exactly equal to the bare mechanical frequency Ω because, when the amplification mechanism is active, the qubit is more often excited during a specific part of the mechanical oscillation, shifting the rest position by $-2x_{\text{zpf}}g_{\text{m}}/\Omega$. Therefore, the MO does not describe a perfect circle in the phase space which alters its apparent frequency. When the thermal noise dominates, this phase evolves randomly whereas, interestingly, its fluctuations are suppressed when the system is in the steady state above thermal noise. This is also visible in the trajectories of the MO in the phase space, in the frame rotating at the effective mechanical frequency, plotted in Fig. 5.5a. Indeed, the complex mechanical amplitude for the large values of θ (in green and red) are more or less contained in a limited area of the phase space defined by the average amplitude and phase.

This decrease in the phase fluctuations is reminiscent of lasers. A laser is a device that amplifies light starting from noise: a pump excites atoms in an amplifying medium inside a cavity, so photons from the incident incoherent light field trigger stimulated emission, eventually leading to the creation of a coherent light field. The gain G_0 of the laser is defined as the factor by which the intensity of the light field is multiplied when going through the amplifying medium once, i.e. an initial intensity I becomes $G_0 I$ after going through the medium. One of the characteristics of lasers is that the gain exhibits a threshold such that if G_0 is below the threshold, there is no amplification while above the threshold, a coherent light field is generated and the intensity is proportional to G_0 [10]. Therefore, above the threshold, the light field has a stabilized amplitude and phase. Another signature of laser behavior is given by the classical second-order correlation function:

$$g^{(2)}(\tau) := \frac{\langle I(t+\tau)I(t)\rangle}{\langle I(t)^2\rangle}, \tag{5.58}$$

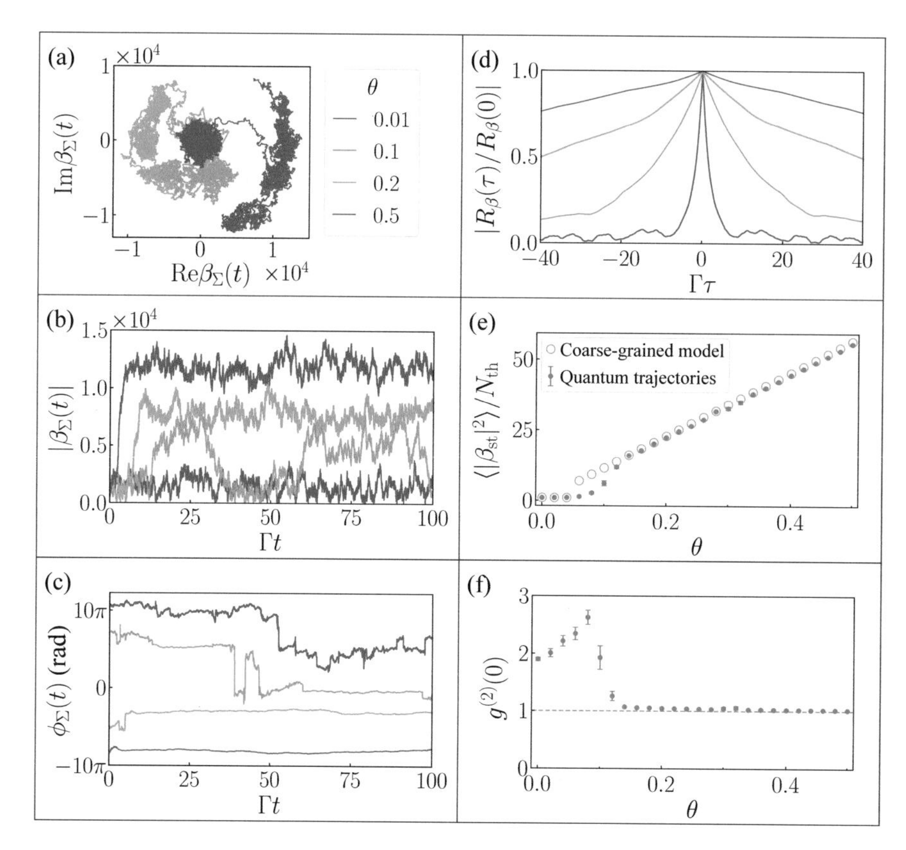

Fig. 5.5 Amplification with a blue-detuned laser. *Left column.* Time evolution of the MO for different values of the excitation rate θ: **a** trajectory in phase space, **b** amplitude and **c** phase. *Right column.* Threshold characterization: **d** autocorrelation of β for different values of θ, **e** average phonon number N_{st} in the steady state as a function of the excitation rate and **f** classical second-order correlation function $g^{(2)}(0)$ as a function of the excitation rate. The legend in **a** also applies to **b**, **c** and **d**. The error bars in **e** and **f** represent the standard error of the mean. *Parameters:* $\Omega/2\pi = 600$ kHz, $g_{\mathrm{m}}/2\pi = 800$ kHz, $\gamma/\Omega = 0.5$, $\Gamma = 20$ Hz, $T = 80$ K and $\Delta/2g_{\mathrm{m}} = 3\sqrt{N_{\mathrm{th}}}$. See Appendix A.2 for more details about the numerical simulations

where I is the intensity of the light. For a laser, $g^{(2)}(0)$ is strictly larger than 1 below the threshold (equals to 2 if the light field is thermal) and goes to 1 above threshold because the light field is coherent. In the case of the hybrid optomechanical system, we have phonons instead of photons and the input field is thermal noise coming from the environment of the MO. We now look for more signatures of laser-like behavior in the steady-state of the MO.

We first evaluate the coherence of the phonon field in the MO through its correlation time. The absolute value of the normalized autocorrelation of β, that is $|R_{\beta}(\tau)/R_{\beta}(0)|$, is represented in Fig. 5.5d. The autocorrelation of β_{Σ} has been approximated by the expression

$$R_{\beta_\Sigma}(\tau) := \int_{t_0}^{t_f} \mathrm{d}t \beta_\Sigma(t)\beta_\Sigma^*(t-\tau). \tag{5.59}$$

The initial time t_0 is chosen large enough to ensure that the amplification mechanism has started and that the MO has reached the steady state with the largest phonon number (except for $\theta = 0.1$ due to the bistability). The final time is such that $t_f - t_0 = 100/\Gamma$. Then, $R_\beta(\tau)$ is obtained by averaging $R_{\beta_\Sigma}(\tau)$ over 16 trajectories. The figure evidences that the correlation time of the MO complex amplitude β increases with θ, therefore when θ is larger than 0.1, the phonon field in the MO is more coherent than thermal noise.

The average phonon number reached by the MO in the steady state, N_{st}, is analogous to the intensity of a light field. Figure 5.5e shows that N_{st} increases with θ and exhibits a threshold like for a laser [10], around $\theta = 0.1$. For the quantum trajectories, it was computed by first averaging $|\beta_\Sigma(t)|^2$ between t_0 and t_f, then taking the mean over 16 trajectories:

$$N_{st} = \left\langle \frac{1}{t_f - t_0} \int_{t_0}^{t_f} \mathrm{d}t |\beta_\Sigma(t)|^2 \right\rangle_{\vec{\Sigma}}. \tag{5.60}$$

The values of the largest fixed point computed from the coarse-grained model (orange circles) are in good agreement with the results obtained with quantum trajectories (blue dots). This agreement is less good around the threshold because the fluctuations neglected in this model matter more around this point.

Furthermore, the classical second-order correlation function also exhibits a laser-like behavior. For the phonon field, it is defined as

$$g^{(2)}(\tau) := \frac{\langle N(t+\tau)N(t)\rangle}{N_{st}^2}, \tag{5.61}$$

with

$$\langle N(t+\tau)N(t)\rangle := \left\langle \frac{1}{t_f - t_0} \int_{t_0}^{t_f} \mathrm{d}t |\beta_\Sigma(t+\tau)|^2 |\beta_\Sigma(t)|^2 \right\rangle_{\vec{\Sigma}}. \tag{5.62}$$

Figure 5.5f shows that $g^{(2)}(0)$ starts from close to 2 in $\theta = 0$, the expected value for a thermal state, then it has an erratic behavior around the threshold and finally goes toward 1, the expected value for a coherent state, when θ increases. The behavior of the MO exhibited in Fig. 5.5d, e, f is consistent with phonon lasing, the excitation rate θ being the equivalent of the laser gain G_0.

5.4.2 *Red-Detuned Laser: Decrease in the Phonon Number*

If we change the sign of the detuning to $\Delta < 0$, the fixed point study of the phonon number N (Fig. 5.4, right column) predicts that the energy conversion process is reversed, so that the mechanical energy can be reduced. These results are confirmed by the numerical simulation of quantum trajectories $\beta_\Sigma(t)$. Indeed, as shown in Fig. 5.6b, when $N_{\mathrm{min}} < N_{\mathrm{th}}$ (in blue), the mechanical amplitude is smaller than the thermal noise (dotted red line) and corresponds to the case represented in Fig. 5.4e. Conversely, when N_{min} is a lot larger than N_{th} (in orange), the qubit never interacts with the laser and there is no decrease in the average phonon number, like in Fig. 5.4f. Therefore the amplitude stays around $\sqrt{N_{\mathrm{th}}}$ and the orange line represents only thermal noise. The decrease in the phonon number is also visible in the trajectory in the phase plane (in blue in Fig. 5.6a) which is almost contained inside the circle of radius $\sqrt{N_{\mathrm{min}}}$. Indeed, when the phonon number exceeds N_{min}, the cooling mechanism is turned on because the qubit starts interacting with the laser. As mentioned before, what we call a cooling mechanism is really a mechanism that reduces mechanical energy, hence the average phonon number.

The absolute value of the normalized autocorrelation of β, $|R_\beta(\tau)/R_\beta(0)|$, is represented in Fig. 5.6c. $R_\beta(\tau)$ is computed in the same way as in the previous part and the initial time t_0 is chosen so that the mechanical amplitude has reached its steady state, i.e. we are not taking into account the initial decrease in $|\beta|$ visible in Fig. 5.6b. The figure shows that the correlation time of the mechanics for $N_{\mathrm{min}} < N_{\mathrm{th}}$ (in blue) is decreased compared to the noise's. For $N_{\mathrm{min}} > N_{\mathrm{th}}$ (in orange), the correlation time is identical to the thermal noise's.

The average phonon number in the steady state N_{st}, given by Eq. (5.60), depends on the detuning, as shown in Fig. 5.6e. For large values of $|\Delta|$, the qubit never interacts with the laser (like in Fig. 5.4f) so N_{st} remains equal to the thermal phonon number. Then, N_{st} decreases with $|\Delta|$ until the detuning gets to close to zero and the cooling efficiency becomes too small to counteract the thermal noise. The predictions of the coarse-grained model (orange circles) have the same general trend as the results obtained with the numerical simulation of quantum trajectories (blue dots). However, the start-up of the cooling mechanism happens for smaller values of $|\Delta|$. This is because this simple model does not take thermal fluctuations into account whereas they can trigger the cooling mechanism.

The classical second-order correlation function $g^{(2)}(0)$, given by Eq. (5.61) is plotted in Fig. 5.6f as a function of the detuning. It is close to 2 for large $|\Delta|$, as expected for a thermal state. Then it slightly decreases with $|\Delta|$ before increasing to reach a maximum around the optimal $|\Delta|$ (the one corresponding to the minimal phonon number). This indicates that the phonon distribution is neither thermal nor coherent. This is confirmed by Fig. 5.6d which shows the probability distribution of $|\beta|^2$ for different detunings. These distributions have a cut-off at $|\beta|^2 = N_{\mathrm{min}}$ which makes this cooling process very similar to evaporative cooling.

5.4.3 Energy Conversion Efficiency

One energy conversion event consists of the absorption by the qubit of one photon of energy $\hbar\omega_{\mathrm{L}}$ from the laser, followed later on by the emission of a photon of energy

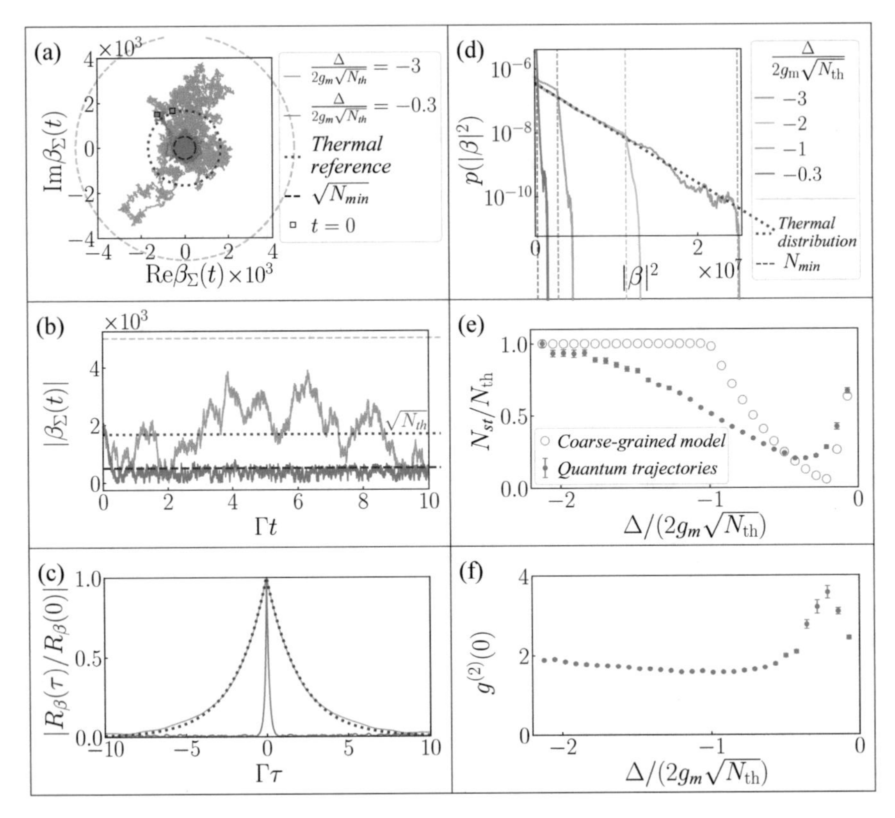

Fig. 5.6 Cooling with a red-detuned laser. Time evolution of the MO for different detunings: **a** trajectory of the MO in phase space and **b** amplitude of the MO. Cooling characterization: **c** absolute value of the normalized autocorrelation of β for different detunings, **d** probability distributions of $|\beta|^2$ for different detunings, **e** average phonon number N_{st} in the steady state as a function of the detuning and **f** classical second-order correlation function $g^{(2)}(0)$ as a function of the detuning. The legend of **a** also applies to **b** and **c**. The error bars in **e** and **f** represent the standard error of the mean. *Parameters:* $\Omega/2\pi = 600$ kHz, $g_{\mathrm{m}}/2\pi = 800$ kHz, $\gamma/\Omega = 0.5$, $\Gamma = 20$ Hz, $T = 80$ K and $\theta = 0.5$, except in **e** and **f** where $\theta = 0.1$ has been chosen to better see the return to the thermal state when the detuning goes to zero. See Appendix A.2 for more details about the numerical simulations

$\hbar\omega_{\mathrm{em}}$, where ω_{em} is the effective frequency of the qubit at the time of the emission. The conversion efficiency of the i-th absorption—emission stochastic event along a trajectory therefore reads

$$\eta_i := \frac{(\hbar\omega_{\mathrm{L}} - \hbar\omega_{\mathrm{em}}^i)}{\hbar\omega_{\mathrm{L}}}, \tag{5.63}$$

With this definition, a negative η_i corresponds to energy taken from the MO. The total energy e_{tot} received by the MO over a time $t_{\mathrm{tot}} = 10/\Gamma$, in units of $\hbar\omega_{\mathrm{L}}$ is obtained by summing these elementary contributions:

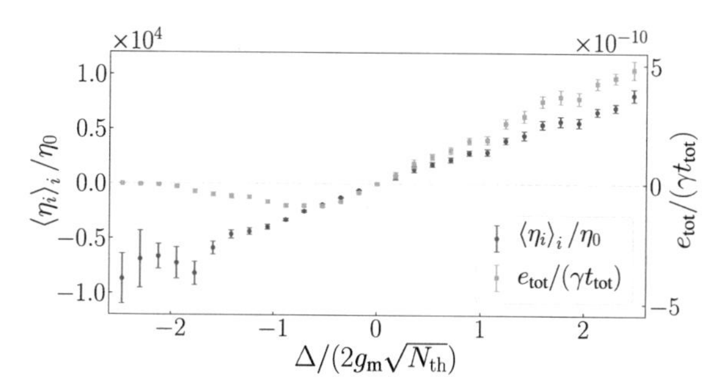

Fig. 5.7 Energy conversion efficiency. *Left axis*: Average conversion efficiency $\langle \eta_i \rangle_i$ normalized to $\eta_0 = \hbar\Omega/\hbar\omega_L$ (in blue) as a function of the detuning. η_i is the fraction of the laser photon's energy received by the MO. It quantifies the efficiency of the energy conversion for the i-th stochastic event: absorption of a laser photon followed by spontaneous emission. *Right axis*: Average energy conversion rate e_{tot}/t_{tot} (in orange) as a function of the laser detuning. e_{tot} is the total energy received by the MO over a time $t_{tot} = 100/\Gamma$ in units of the laser energy. The error bars in both plots represent the standard error of the mean. *Parameters:* $t_{tot}\Gamma = 10$, $\omega_0/2\pi = 500$ THz, $\theta = 0.2$, $\Omega/2\pi = 600$ kHz, $g_m/2\pi = 800$ kHz, $\gamma/\Omega = 0.5$, $\Gamma = 20$ Hz and $T = 80$ K

$$e_{tot} = \sum_i \eta_i. \tag{5.64}$$

The average energy conversion rate e_{tot}/t_{tot} is plotted in Fig. 5.7 (orange squares) as a function of the detuning. For red detunings, e_{tot} is negative since the MO is being cooled down and $-e_{tot}/t_{tot}$ corresponds to the average cooling power in units of $\hbar\omega_L$. The same drop in the energy conversion rate as in Fig. 5.6e is noticed when the detuning goes to zero. As expected, when the detuning is zero, the energy conversion rate goes to zero since there are as many cooling events as amplifying ones. Finally, for blue detunings, the average energy conversion rate is positive and increase with the detuning. However, this increase in the average energy conversion, as well as the increase in the phonon number in the steady state) comes with a trade-off: The larger the detuning, the longer the average amplification start-up time (i.e. the first time the noise makes the phonon number greater than N_{min}).

The conversion efficiency, averaged over the stochastic events and normalized to

$$\eta_0 := \frac{\hbar\Omega}{\hbar\omega_L}, \tag{5.65}$$

is represented in Fig. 5.7 (blue dots) as a function of the detuning. η_0 corresponds to the efficiency of the resolved-sideband cooling scheme in which each photon that is absorbed by the lower sideband removes one quantum of energy from the MO [16]. Note that we are not in the resolved-sideband regime $\gamma \ll \Omega$ but in the regime $\gamma \sim \Omega$. The figure shows that the average conversion efficiency of a single event is orders of magnitude larger than the one of resolved-sideband cooling. Unlike the energy conversion rate, the conversion efficiency does not go to zero for large negative values of Δ. Indeed, though cooling events are really rare in this case, they are very efficient and the conversion efficiency does not take into account their rarity, which accounts for the larger error bars.

5.5 Summary

In this chapter, we evidenced that hybrid optomechanical systems can be seen as autonomous machines operating between two baths. The qubit corresponds to the working substance, the MO plays the role of the battery, the cold bath is a thermal bath at zero temperature and the hot bath is a "colored" bath (a monochromatic laser). We can choose whether the MO provides or extract work via the sign of the detuning between the laser and the qubit.

We first derived the master equation for the optomechanical system, taking into account the thermal baths of both the qubit and the mechanical oscillator (MO), in the regime $\omega_0 \gg \gamma, g_{\mathrm{m}}, \Omega \gg \Gamma$. Then, we unraveled this equation into quantum trajectories. The interaction between the laser and the qubit was modeled in an effective way and added in the trajectories in the form of a probability θ that the qubit ends in the excited state after being in resonance with the laser. Finally, we were able to simulate numerically the stochastic evolution of the MO which always remains in a coherent state.

We also elaborated a simpler model by coarse-graining the evolution of the phonon number in the MO over one mechanical period. This model allowed us to determine that the optimal value of the spontaneous emission rate of the qubit is $\gamma \simeq \Omega/2$. We also studied the fixed points of the phonon number, estimating the average phonon number in the steady states, and determined the interesting regimes for energy conversion. This study showed that hybrid optomechanical systems can be considered as autonomous and reversible thermal machines allowing to amplify or reduce the mechanical motion. The direction of the energy conversion is determined by the sign of the detuning between the laser and the frequency of the bare qubit. When the laser is blue-detuned, the qubit absorbs a high energy photon from the laser and emits a lower energy one, the energy difference being given to the MO. When the laser is red-detuned, the qubit absorbs a low energy photon from the laser and emits a higher energy one, the energy difference being provided by the MO.

Then, we studied the evolution of the MO using numerically generated quantum trajectories with experimentally realistic parameters. We evidenced that, with a blue detuning, if the excitation rate θ is large enough, a coherent phonon state is built starting from thermal noise. The behavior of the MO in this regime exhibits several signatures of phonon lasing, including a threshold in θ, which plays the role of the gain. However, unlike in usual lasers, no population inversion is required and the system is bistable, so if θ is not a lot larger than the threshold, thermal fluctuations can bring back the MO in its thermal state fixed point. Conversely, with a red detuning, the average phonon number can be reduced below the thermal number and this process is similar to evaporative cooling, with a cut-off for mechanical amplitudes larger than the threshold amplitude triggering the cooling mechanism. Moreover, on a single cooling event, this process is orders of magnitude more efficient than resolved-sideband cooling.

As a perspective, it would be interesting to investigate further the cooling process, and especially determine whether the MO eventually thermalizes. We could also study the entropy of the MO, as cooling in the thermodynamic sense is associated with a decrease in entropy. Finally, it would be interesting to determine the cooling limit, which would require a more detailed modeling of the interaction between the qubit and the laser.

References

1. Scovil HED, Schulz-DuBois EO (1959) Three-level masers as heat engines. Phys Rev Lett 2(6):262–263. https://doi.org/10.1103/physrevlett.2.262
2. Ghosh A, Gelbwaser-Klimovsky D, Niedenzu W, Lvovsky AI, Mazets I, Scully MO, Kurizki G (2018) Two-level masers as heat-to-work converters. Proc Natl Acad Sci USA 115(40):9941–9944. https://doi.org/10.1073/pnas.1805354115
3. Jing H, Özdemir SK, Lü X-Y, Zhang J, Yang L, Nori F (2014) PT-Symmetric phonon laser. Phys Rev Lett 113(5):053604. https://doi.org/10.1103/PhysRevLett.113.053604
4. Jiang Y, Maayani S, Carmon T, Nori F, Jing H (2018) Nonreciprocal phonon laser. Phys Rev Appl 10(6):064037. https://doi.org/10.1103/PhysRevApplied.10.064037
5. Zhang Y-L, Zou C-L, Yang C-S, Jing H, Dong C-H, Guo G-C, Zou X-B (2018) Phase-controlled phonon laser. New J Phys 20(9):093005. https://doi.org/10.1088/1367-2630/aadc9f
6. Kabuss J, Carmele A, Brandes T, Knorr A (2012) Optically driven quantum dots as source of coherent cavity phonons: a proposal for a phonon laser scheme. Phys Rev Lett 109(5):054301. https://doi.org/10.1103/PhysRevLett.109.054301
7. Elouard C, Richard M, Auffèves A (2015) Reversible work extraction in a hybrid opto-mechanical system. New J Phys 17(5):055018. https://doi.org/10.1088/1367-2630/17/5/055018
8. Cohen-Tannoudji C, Dupont-Roc J, Grynberg G (2004) Atom-photon interactions: basic processes and applications, ser. Physics textbook. Wiley, Weinheim-VCH
9. Gardiner C, Zoller P (2004) Quantum noise: a handbook of Markovian and Non-Markovian quantum stochastic methods with applications to quantum optics, ser. Springer series in synergetics. Springer, Berlin, Heidelberg
10. Grynberg G, Aspect A, Fabre C (2010) Introduction to quantum optics: from the semi-classical approach to quantized light. Cambridge University Press, Cambridge
11. Auffèves A, Richard M (2014) Optical driving of macroscopic mechanical motion by a single two-level system. Phys Rev A 90(2):023818. https://doi.org/10.1103/PhysRevA.90.023818
12. Pirkkalainen J-M, Cho SU, Li J, Paraoanu GS, Hakonen PJ, Sillanpää MA (2013) Hybrid circuit cavity quantum electrodynamics with a micromechanical resonator. Nature 494(7436):211–215. https://doi.org/10.1038/nature11821
13. Yeo I, de Assis P-L, Gloppe A, Dupont-Ferrier E, Verlot P, Malik NS, Dupuy E, Claudon J, Gérard J-M, Auffèves A, Nogues G, Seidelin S, Poizat J-P, Arcizet O, Richard M (2014) Strain-mediated coupling in a quantum Dot-Mechanical oscillator hybrid system. Nat Nanotech 9:106–110. https://doi.org/10.1038/nnano.2013.274
14. Shamirzaev TS, Debus J, Abramkin DS, Dunker D, Yakovlev DR, Dmitriev DV, Gutakovskii AK, Braginsky LS, Zhuravlev KS, Bayer M (2011) Exciton recombination dynamics in an ensemble of (In,Al)As/AlAs quantum dots with indirect band-gap and type-I band alignment. Phys Rev B 84(15):155318. https://doi.org/10.1103/PhysRevB.84.155318

15. Debus J, Shamirzaev TS, Dunker D, Sapega VF, Ivchenko EL, Yakovlev DR, Toropov AI, Bayer M (2014) Spin-flip Raman scattering of the Γ-X mixed exciton in indirect band gap (In,Al)As/AlAs quantum dots. Phys Rev B 90(12):125431. https://doi.org/10.1103/PhysRevB.90.125431
16. Aspelmeyer M, Kippenberg TJ, Marquardt F (2014) Cavity optomechanics. Rev Mod Phys 86(4):1391–1452. https://doi.org/10.1103/RevModPhys.86.1391

Chapter 6
Coherent Quantum Engine

Coherence plays a key role in quantum information, which is why it has been considered as a potential resource for quantum machines, with the aim of surpassing classical ones [1–6]. In Refs. [5, 6], the quantum coherence in the working substance is injected by the drive while in Ref. [1] it comes from the bath which is non-thermal. However there has been no experimental implementation of such a quantum machine using a single qubit as working substance so far.

Until now, we have only studied cases where no coherence in the qubit energy eigenbasis was ever involved. While Chaps. 3–5 focused on hybrid optomechanical systems where the MO acts as a battery dispersively coupled to the qubit, in this chapter, we consider a resonant battery that coherently drives the qubit. This situation is particularly suitable for the study of the impact of coherence in quantum machines.

We use an engineered bath to prepare the qubit in an arbitrary superposition of energy eigenstates. This bath acts as a source of energy and coherence, allowing us to make a two-stroke quantum engine extracting work from a single bath. As the bath is non-thermal, this engine does not violate the laws of thermodynamics. It consists of a resonantly driven qubit embedded in a waveguide, usually called "one-dimensional atom" and can be implemented with state-of-the-art artificial atoms coupled to superconducting [7, 8] or semiconducting circuits [9, 10].

We derive the evolution of the state of the qubit in contact first with the engineered bath and secondly with the battery and show that these two situations can be combined to create a two-stroke quantum engine. Then, we focus on the regime of strong driving where the battery is loaded with a large number of photons, which corresponds to classical Rabi oscillations, and evidence that coherence boosts the engine's power. Finally, we study arbitrary driving strengths down to the spontaneous regime where the battery is not loaded. In this latter regime, coherence determines the amount of energy coherently emitted in the waveguide. This study is presented in [11].

J. Monsel, *Quantum Thermodynamics and Optomechanics*, Springer Theses,
https://doi.org/10.1007/978-3-030-54971-8_6

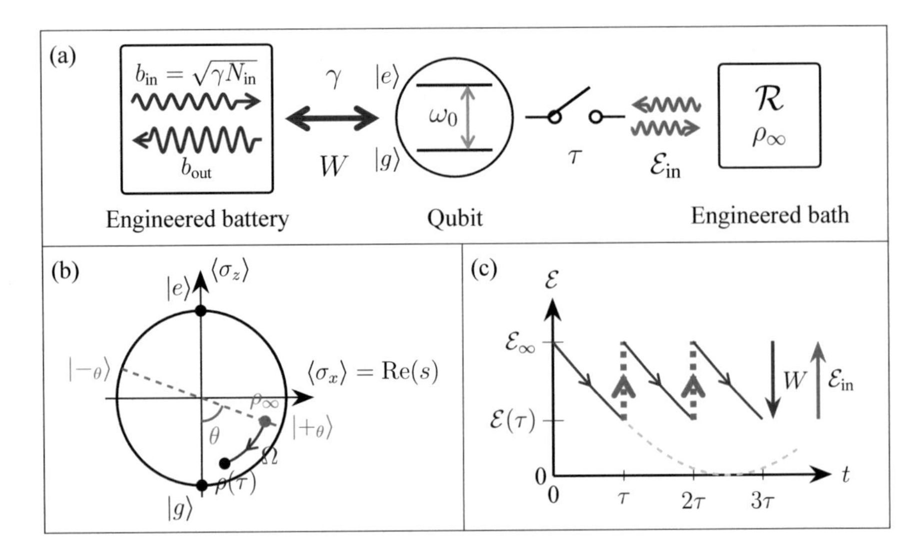

Fig. 6.1 Engine's schematic. **a** Situation under study: a qubit of transition frequency ω_0 is coupled to a waveguide. γ denotes the spontaneous emission rate of the qubit, b_{in} the mean amplitude of the coherent input drive, γN_{in} the photon input rate and b_{out} the mean amplitude of the output field. Work corresponds to the coherent fraction of the energy radiated by the qubit in the waveguide. The modes of the waveguide thus play the role of the battery whose load is defined by N_{in}. **b** and **c** Two-stroke engine when the battery is classical ($N_{in} \gg 1$). **b** Evolution of the state of the qubit in the Bloch sphere. *First stroke:* At time $t = 0$, the qubit is put in contact with the engineered bath that prepares it in ρ_∞ of eigenstates $|\pm_\theta\rangle$ and energy ε_∞. *Second stroke:* For $t \in [0, \tau]$, the bath is decoupled and the qubit unitarily evolves from ρ_∞ to $\rho(\tau)$ of energy $\varepsilon(\tau)$. **c** Time evolution of the energy of the qubit. During the second stroke, an amount of work $W = \varepsilon(\tau) - \varepsilon_\infty$ is extracted in the battery, that corresponds to the energy $\mathcal{E}_{in}$ provided by the engineered bath

6.1 Microscopic Description of the Engine

The setup under study, represented in Fig. 6.1a, is a qubit embedded in a one-dimensional waveguide. This qubit can be, on the one hand, coupled to an engineered bath and, on the other hand, coherently driven by an engineered battery. This battery corresponds to the modes of the waveguide resonant with the qubit's transition. We derive separately the evolution of the qubit first in contact with the engineered bath, secondly when driven by the battery. Then, we combine the two to make a two-stroke engine.

6.1.1 Engineered Bath: Modeling and Interaction

Reservoir engineering techniques [12] allow the preparation of arbitrary quantum states by non-unitary operations. This ability to shape dissipation provides new kinds of baths, opening the way for various applications in quantum thermodynamics, like measuring stochastic heat exchanges [13], creating steady-state entanglement [14] or charging a quantum battery [15]. In this chapter, we use an engineered bath to prepare the qubit in a coherent superpositions of energy eigenstates during the engine's first stroke. This bath consists of a pump, which is quasi-resonant with the qubit's frequency, and an electromagnetic reservoir whose mode density is engineered, for example, by using a cavity [16] or a photonic crystal [17]. We first derive the master equation describing the evolution of the qubit when coupled to this bath. Then, we explain how an arbitrary state can be prepared.

6.1.1.1 Derivation of the Master Equation for the Driven Qubit

The qubit is coupled both to a quasi-resonant pump at frequency ω_{P}, and whose amplitude produces the Rabi frequency Ω_{R}, and to a reservoir of electromagnetic modes at temperature T. Note that this pump is much stronger than the input drive used as a battery. The evolution of the total system is governed by the Hamiltonian

$$H(t) = H_{\mathrm{q}}^{\mathrm{P}}(t) + H_{\mathcal{R}} + V \tag{6.1}$$

where

$$H_{\mathrm{q}}^{\mathrm{P}}(t) = H_0 - \frac{\hbar\Omega_{\mathrm{R}}}{2}(\mathrm{e}^{\mathrm{i}\omega_{\mathrm{P}}t}\sigma_- + \mathrm{e}^{-\mathrm{i}\omega_{\mathrm{P}}t}\sigma_+) \tag{6.2}$$

is the qubit's Hamiltonian, $H_{\mathcal{R}}$ (Eq. (3.13)) is the Hamiltonian of the reservoir and $V = \sum_{l=\pm} R_l\sigma_l$ (Eq. (3.15)) is the coupling Hamiltonian in the rotating wave approximation. We have denoted $H_0 = \hbar\omega_0|e\rangle\langle e|$, the bare Hamiltonian of the qubit. In the frame rotating at the frequency ω_{P}, the Hamiltonian becomes

$$\tilde{H} = \tilde{H}_{\mathrm{q}}^{\mathrm{P}}(t) + H_{\mathcal{R}} + \tilde{V}(t), \tag{6.3}$$

with

$$\tilde{H}_{\mathrm{q}}^{\mathrm{P}}(t) = \hbar\Delta|e\rangle\langle e| - \frac{\hbar\Omega_{\mathrm{R}}}{2}\sigma_x, \tag{6.4}$$

$$\tilde{V}(t) = \sum_{l=\pm} R_l\sigma_l\mathrm{e}^{\mathrm{i}l\omega_{\mathrm{P}}t}. \tag{6.5}$$

We have defined

$$\Delta := \omega_0 - \omega_{\mathrm{P}}, \tag{6.6}$$

the pump detuning and $\sigma_x := |e\rangle\langle e|g + |g\rangle\langle g|e$. $\tilde{H}_{\mathrm{q}}^{\mathrm{P}}$ can be rewritten in the form

$$\tilde{H}_{\mathrm{q}}^{\mathrm{P}} = \frac{\hbar\Omega_{\mathrm{P}}}{2}\Sigma_z^{\theta} + \frac{\hbar\Delta}{2}, \tag{6.7}$$

where

$$\Omega_{\rm P} := \sqrt{\Omega_{\rm R}^2 + \Delta^2} \tag{6.8}$$

is the generalized Rabi frequency and

$$\Sigma_z^\theta := |-_\theta\rangle\langle -_\theta| - |+_\theta\rangle\langle +_\theta|. \tag{6.9}$$

The two eigenstates of $\tilde{H}_{\rm q}^{\rm P}$ reads

$$|+_\theta\rangle := \sin\left(\frac{\theta}{2}\right)|e\rangle + \cos\left(\frac{\theta}{2}\right)|g\rangle, \tag{6.10a}$$

$$|-_\theta\rangle := -\cos\left(\frac{\theta}{2}\right)|e\rangle + \sin\left(\frac{\theta}{2}\right)|g\rangle, \tag{6.10b}$$

with $\theta = \arctan(\Omega_{\rm R}/\Delta)$ (See Fig. 6.1b).

Following [18, 19] and applying the Born-Markov approximation, in the interaction picture, the precursor of the qubit's master equation reads

$$\begin{aligned}\Delta\rho^{\rm I}(t) &= \rho^{\rm I}(t+\Delta t) - \rho^{\rm I}(t)\\ &= -\frac{1}{\hbar^2}\int_t^{t+\Delta t} {\rm d}t' \int_0^\infty {\rm d}\tau {\rm Tr}_{\mathcal{R}}\left(\left[\tilde{V}^{\rm I}(t'), \left[\tilde{V}^{\rm I}(t'-\tau), \rho^{\rm I}(t)\otimes\rho_{\mathcal{R}}\right]\right]\right),\end{aligned} \tag{6.11}$$

We have chosen the time step Δt such that

$$\tau_{\rm c}, \omega_0^{-1}, \omega_{\rm P}^{-1}, \Omega_{\rm R}^{-1} \ll \Delta t \ll \gamma^{-1}. \tag{6.12}$$

$\rho(t)$ is the density operator of the qubit and the exponent I denotes the interaction picture such that

$$\begin{aligned}\tilde{V}^{\rm I}(t) &= {\rm e}^{\frac{\rm i}{\hbar}(\tilde{H}_{\rm q}^{\rm P}+H_{\mathcal{R}})t}\,\tilde{V}(t){\rm e}^{-\frac{\rm i}{\hbar}(\tilde{H}_{\rm q}^{\rm P}+H_{\mathcal{R}})t}\\ &= \sum_{l=\pm} R_l^{\rm I}(t)\sigma_l^{\rm I}(t){\rm e}^{{\rm i}l\omega_{\rm P}t},\end{aligned} \tag{6.13}$$

and

$$R_l^{\rm I}(t) = \sum_k \hbar g_k a_k {\rm e}^{-l{\rm i}\omega_k t}, \tag{6.14}$$

$$\sigma_l^{\rm I}(t) = \exp\left(\frac{\rm i}{\hbar}\tilde{H}_{\rm q}^{\rm P}t\right)\sigma_l \exp\left(-\frac{\rm i}{\hbar}\tilde{H}_{\rm q}^{\rm P}t\right). \tag{6.15}$$

The trace over the reservoir gives terms of the form $g_{ll'}(u,v) = \frac{1}{\hbar^2}\mathrm{Tr}_{\mathcal{R}}((R_l^{\mathrm{I}}(u))^\dagger R_{l'}^{\mathrm{I}}(v)\rho_{\mathcal{R}})$, which are zero if $l \neq l'$, therefore

$$\Delta\rho^{\mathrm{I}}(t) = -\int_t^{t+\Delta t} \mathrm{d}t' \int_0^\infty \mathrm{d}\tau \sum_{l=\pm} g_{ll}(t', t'-\tau)\Big((\sigma_l^{\mathrm{I}}(t'))^\dagger \sigma_l^{\mathrm{I}}(t'-\tau)\rho^{\mathrm{I}}(t) - \sigma_l^{\mathrm{I}}(t'-\tau)\rho^{\mathrm{I}}(t)(\sigma_l^{\mathrm{I}}(t'))^\dagger\Big) + \text{h.c.} \tag{6.16}$$

Then, we use the decomposition

$$\sigma_l^{\mathrm{I}}(t) = \sum_{\omega=0,\pm\Omega_{\mathrm{P}}} \tilde{\sigma}_l(\omega)\mathrm{e}^{-\mathrm{i}\omega t}, \tag{6.17}$$

where we have defined

$$\tilde{\sigma}_\pm(0) := -\frac{\sin(\theta)}{2}\Sigma_z^\theta \tag{6.18a}$$

$$\tilde{\sigma}_\pm(-\Omega_{\mathrm{P}}) := -\frac{\pm 1 + \cos(\theta)}{2}\Sigma_-^\theta, \tag{6.18b}$$

$$\tilde{\sigma}_\pm(\Omega_{\mathrm{P}}) := -\frac{\mp 1 + \cos(\theta)}{2}\Sigma_+^\theta, \tag{6.18c}$$

and $\Sigma_-^\theta := |-_\theta\rangle\langle -_\theta| +_\theta$, $\Sigma_+^\theta := |+_\theta\rangle\langle +_\theta| -_\theta$. Therefore, Eq. (6.16) becomes

$$\Delta\rho^{\mathrm{I}}(t) = -\int_t^{t+\Delta t} \mathrm{d}t' \sum_{l=\pm}\sum_{\omega,\omega'} \mathrm{e}^{\mathrm{i}(\omega-\omega')t'} \int_0^\infty \mathrm{d}\tau\, g_{ll}(t', t'-\tau)\mathrm{e}^{\mathrm{i}(\omega'-l\omega_{\mathrm{P}})\tau} \times \left(\tilde{\sigma}_l^\dagger(\omega)\tilde{\sigma}_l(\omega')\rho^{\mathrm{I}}(t) - \tilde{\sigma}_l(\omega')\rho^{\mathrm{I}}(t)\tilde{\sigma}_l^\dagger(\omega)\right) + \text{h.c.} \tag{6.19}$$

$g_{ll}(t', t'-\tau)$ only depends on τ and

$$\int_0^\infty \mathrm{d}\tau\, g_{ll}(t', t'-\tau)\mathrm{e}^{\mathrm{i}(\omega'-l\omega_{\mathrm{P}})\tau} = \frac{1}{2}\gamma(\omega_{\mathrm{P}} - l\omega')(\bar{n}(\omega_{\mathrm{P}} - l\omega') + \delta_{l-}). \tag{6.20}$$

$\gamma(\omega) := \gamma_\omega$ (Eq. (3.16)) is the spontaneous emission rate of the qubit at frequency ω and $\bar{n}(\omega) := \bar{n}_\omega$ (Eq. (3.14)) is the mean number of photons at the frequency ω in the reservoir. To put the master equation in the Lindblad form we use the secular approximation [20], that consists in neglecting the terms evolving in $\omega - \omega' \neq 0$. Indeed, integrating them over t' gives terms in $\mathrm{sinc}((\omega-\omega')\Delta t/2)$ and $|\omega - \omega'|\Delta t \geq \Omega_{\mathrm{P}}\Delta t \gg 1$. Finally, we obtain

$$\dot{\rho}^{\mathrm{I}}(t) = (\mathcal{L}_0 + \mathcal{L}_1 + \mathcal{L}_2)\rho^{\mathrm{I}}(t), \tag{6.21}$$

with

$$\mathcal{L}_0 = (\gamma_{0\uparrow} + \gamma_{0\downarrow})D[\Sigma_z^\theta], \tag{6.22}$$
$$\mathcal{L}_i = \gamma_{i\uparrow} D[\Sigma_+^\theta] + \gamma_{i\downarrow} D[\Sigma_-^\theta],\ i = 1, 2. \tag{6.23}$$

$\mathcal{L}_0$ corresponds to the pure-dephasing in the dressed basis, with the rates

$$\gamma_{0\uparrow} = \frac{\sin^2(\theta)}{4}\gamma(\omega_{\mathrm{P}})\bar{n}(\omega_{\mathrm{P}}), \tag{6.24}$$
$$\gamma_{0\downarrow} = \frac{\sin^2(\theta)}{4}\gamma(\omega_{\mathrm{P}})(\bar{n}(\omega_{\mathrm{P}}) + 1). \tag{6.25}$$

$\mathcal{L}_1$ and $\mathcal{L}_2$ correspond to thermal relaxation with respective rates

$$\gamma_{1\uparrow} = \cos^4\left(\frac{\theta}{2}\right)\gamma(\omega_{\mathrm{P}} + \Omega_{\mathrm{P}})(\bar{n}(\omega_{\mathrm{P}} + \Omega_{\mathrm{P}}) + 1), \tag{6.26}$$
$$\gamma_{1\downarrow} = \cos^4\left(\frac{\theta}{2}\right)\gamma(\omega_{\mathrm{P}} + \Omega_{\mathrm{P}})\bar{n}(\omega_{\mathrm{P}} + \Omega_{\mathrm{P}}), \tag{6.27}$$
$$\gamma_{2\uparrow} = \sin^4\left(\frac{\theta}{2}\right)\gamma(\omega_{\mathrm{P}} - \Omega_{\mathrm{P}})\bar{n}(\omega_{\mathrm{P}} - \Omega_{\mathrm{P}}), \tag{6.28}$$
$$\gamma_{2\downarrow} = \sin^4\left(\frac{\theta}{2}\right)\gamma(\omega_{\mathrm{P}} - \Omega_{\mathrm{P}})(\bar{n}(\omega_{\mathrm{P}} - \Omega_{\mathrm{P}}) + 1). \tag{6.29}$$

Equation (6.21) can be interpreted in the radiative cascade picture [20], modeling the pump as a quantized field. The eigenstates of the qubit-pump system read

$$|+_\theta(n_{\mathrm{P}})\rangle = \sin\left(\frac{\theta}{2}\right)|e, n_{\mathrm{P}} - 1\rangle + \cos\left(\frac{\theta}{2}\right)|g, n_{\mathrm{P}}\rangle, \tag{6.30a}$$

$$|-_\theta(n_{\mathrm{P}})\rangle = -\cos\left(\frac{\theta}{2}\right)|e, n_{\mathrm{P}} - 1\rangle + \sin\left(\frac{\theta}{2}\right)|g, n_{\mathrm{P}}\rangle, \tag{6.30b}$$

where n_{P} is the number of photons in the field. These states form the Jaynes-Cummings ladder, depicted in Fig. 6.2a. The frequency of the transition between $|+_\theta(n_{\mathrm{P}})\rangle$ and $|-_\theta(n_{\mathrm{P}})\rangle$ scales like $\sqrt{n_{\mathrm{P}}}$. In the classical limit considered here, the pump is a coherent field $|\alpha_{\mathrm{P}}\rangle$ containing a large number of photons $\bar{n}_{\mathrm{P}} = |\alpha_{\mathrm{P}}|^2 \gg 1$. Therefore $\sqrt{\bar{n}_{\mathrm{P}} + 1} \simeq \sqrt{\bar{n}_{\mathrm{P}}}$ and the levels in the ladder can be considered as equally spaced. Therefore, the transition $|-_\theta\rangle \to |+_\theta\rangle$, associated with the jump operator Σ_+^θ, consists of an ensemble of transitions $|-_\theta(n_{\mathrm{P}})\rangle \to |+_\theta(n_{\mathrm{P}} - 1)\rangle$. It is characterized by the spontaneous emission of a blue-shifted photon of frequency

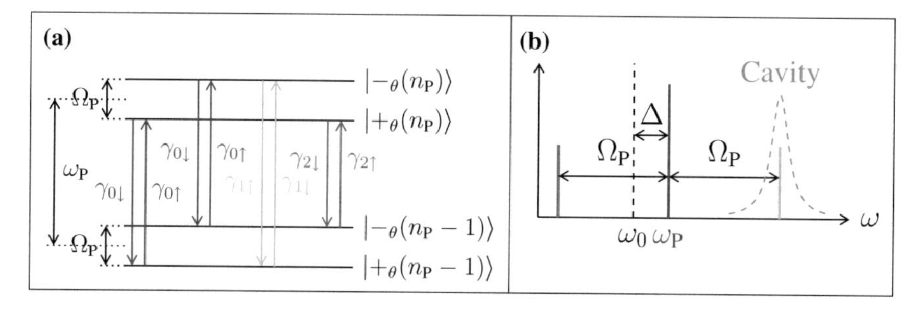

Fig. 6.2 **a** Jaynes-Cummings ladder formed by the eigenstates of the qubit-pump system when the pump field is quantized. **b** Emission spectrum of the qubit in the classical limit where the pump is a coherent field containing a large number of photons. The position of the cavity used to enhance the transition rate $\gamma(\omega_P + \Omega_P)$ is indicated by a gray dashed line

$\omega_P + \Omega_P$. Similarly, the transition $|+_\theta\rangle \to |-_\theta\rangle$ consists of an ensemble of transitions $|+_\theta(n_P)\rangle \to |-_\theta(n_P-1)\rangle$ and is characterized by the emission of a red-shifted photon of frequency $\omega_P - \Omega_P$. Pure dephasing is induced by the transitions $|\pm_\theta(n)\rangle \to |\pm_\theta(n-1)\rangle$ and signaled by photons emitted at the pump frequency. The emitted photons build a spectrum forming the Mollow triplet represented in Fig. 6.2b.

6.1.1.2 Arbitrary State Preparation

In the rest of this chapter, we will assume that the reservoir is at zero temperature, so $\bar{n}(\omega) = 0$ for any ω. The master equation (6.21) thus becomes, in the Schrödinger picture,

$$\dot{\rho}(t) = -\frac{\mathrm{i}}{\hbar}[\tilde{H}_{\mathrm{q}}^{\mathrm{P}}, \rho(t)] + \frac{\cos^2(\theta)}{4}\gamma(\omega_P) D[\Sigma_z^\theta]\rho(t) + \cos^4\left(\frac{\theta}{2}\right)\gamma(\omega_P + \Omega_P) D[\Sigma_+^\theta]\rho(t) + \sin^4\left(\frac{\theta}{2}\right)\gamma(\omega_P - \Omega_P) D[\Sigma_-^\theta]\rho(t). \tag{6.31}$$

The equilibrium state corresponding to this equation is

$$\rho_\infty = p|-_\theta\rangle\langle-_\theta| + (1-p)|+_\theta\rangle\langle+_\theta|, \tag{6.32}$$

where

$$p = \frac{\sin^4(\frac{\theta}{2})\gamma(\omega_P - \Omega_P)}{\sin^4(\frac{\theta}{2})\gamma(\omega_P - \Omega_P) + \cos^4(\frac{\theta}{2})\gamma(\omega_P + \Omega_P)}. \tag{6.33}$$

It is therefore possible to tune $p \in [0, 1/2]$ by changing the transitions rates $\gamma(\omega_P \pm \Omega_P)$. One possibility is to use the Purcell effect by putting a narrow cavity resonant with the $\omega_P + \Omega_P$ transition [16], as represented Fig. 6.1b. Another option is to engineer the density of modes in the environment, for instance with a photonic crystal [17]. Furthermore, the angle $\theta \in [0, \pi]$ fixing the relaxation basis of the engineered bath can be tuned by playing on the pump detuning.

This engineering of modes also greatly enhances the relaxation time of the qubit in the bath, which is given by the fastest transition rate $\gamma(\omega_P \pm \Omega_P)$. Therefore, in this chapter we will neglect the relaxation time, namely the duration of the first stroke, compared to the duration τ of the work extraction.

To summarize, we have shown that we can control both the position of the relaxation basis $\{|-_\theta\rangle, |+_\theta\rangle\}$ on the Bloch sphere and the purity of the state ρ_∞ of the qubit. Besides, this bath engineering strategy is compatible with the properties of the one-dimensional atom, which solely requires the reservoir of modes to be flat around the transition frequency of the qubit. It was experimentally implemented in Ref. [17].

6.1.2 Engineered Battery: Modeling and Interaction

6.1.2.1 Derivation of the Evolution of the Qubit

During the second stroke, the engineered bath is decoupled from the qubit. Therefore the qubit only interacts with a one-dimensional reservoir of electromagnetic modes indexed, in the continuum limit, by their frequency ω and characterized by the normalized density of modes $\rho(\omega)$. We will use the input-output formalism to describe the field propagating in the waveguide [21, 22]. The total Hamiltonian reads

$$H = H_0 + H_{\mathcal{R}_{\mathrm{wg}}} + V_{\mathcal{R}_{\mathrm{wg}}}, \tag{6.34}$$

where

$$H_{\mathcal{R}_{\mathrm{wg}}} := \int_0^\infty \mathrm{d}\omega \hbar\omega \rho(\omega) \hat{b}_\omega^\dagger \hat{b}_\omega, \tag{6.35}$$

$$V_{\mathcal{R}_{\mathrm{wg}}} := \mathrm{i} \int_0^\infty \mathrm{d}\omega \rho(\omega) \hbar g(\omega) (b_\omega^\dagger + b_\omega)(\sigma_- - \sigma_+), \tag{6.36}$$

are respectively the bare Hamiltonian of the qubit, the bare Hamiltonian of the one-dimensional reservoir and the coupling Hamiltonian. We have denoted b_ω the annihilation operator of the mode of frequency ω and $g(\omega)$ the coupling strength between this mode and the qubit. In the rotating wave approximation, $V_{\mathcal{R}_{\mathrm{wg}}}$ can be rewritten

$$V_{\mathcal{R}_{\mathrm{wg}}} = \mathrm{i} \int_{-\infty}^\infty \mathrm{d}\omega \rho(\omega) \hbar g(\omega) (b_\omega^\dagger \sigma_- - \sigma_+ b_\omega). \tag{6.37}$$

Only the quasi-resonant terms contribute significantly to the evolution of the system, so we have dropped the rapidly oscillating terms in $b_\omega^\dagger\sigma_+$ and $b_\omega\sigma_-$. Similarly, we have extended the integral down to $-\infty$ because the added terms are all non resonant, but this will allow mathematical simplifications [22].

In the Heisenberg picture, the equations of evolution of $b_\omega(t)$ and any observable $O_\mathrm{q}(t)$ of the qubit are

$$\dot{b}_\omega(t) = -\,\mathrm{i}\omega b_\omega(t) + g(\omega)\sigma_-(t), \tag{6.38}$$

$$\begin{aligned}\dot{O}_\mathrm{q}(t) = &-\frac{\mathrm{i}}{\hbar}[O_\mathrm{q}(t), H_0(t)]\\ &+\int_{-\infty}^{\infty}\mathrm{d}\omega\rho(\omega)g(\omega)\big(b_\omega^\dagger(t)[O_\mathrm{q}(t), \sigma_-(t)] - [O_\mathrm{q}(t), \sigma_+(t)]b_\omega(t)\big).\end{aligned} \tag{6.39}$$

Formally integrating Eq. (6.38) yields, for any time $t \geq 0$,

$$b_\omega(t) = b_\omega(0)\mathrm{e}^{-\mathrm{i}\omega t} + g(\omega)\int_0^t \mathrm{d}t'\sigma_-(t')\mathrm{e}^{-\mathrm{i}\omega(t-t')}. \tag{6.40}$$

Injecting this result in Eq. (6.39), we obtain

$$\dot{O}_\mathrm{q}(t) = -\frac{\mathrm{i}}{\hbar}[O_\mathrm{q}(t), H_0(t)] - [O_\mathrm{q}(t), \sigma_+(t)]\left(\frac{\gamma}{2}\sigma_-(t) + \sqrt{\gamma}\hat{b}_\mathrm{in}(t)\right) \tag{6.41}$$

$$+\left(\frac{\gamma}{2}\sigma_+(t) + \sqrt{\gamma}\hat{b}_\mathrm{in}^\dagger(t)\right)[O(t), \sigma_-(t)]. \tag{6.42}$$

We have denoted $\hat{b}_\mathrm{in}(t)$ the input operator, defined by

$$\hat{b}_\mathrm{in}(t) := \frac{1}{\sqrt{2\pi}}\int_{-\infty}^{\infty}\mathrm{d}\omega\sqrt{\rho(\omega)}b_\omega(0)\mathrm{e}^{-\mathrm{i}\omega t}, \tag{6.43}$$

which describes the field at time t in the waveguide before its interaction with the qubit, and

$$\gamma := 2\pi g^2(\omega_0)\rho(\omega_0), \tag{6.44}$$

which is the damping rate undergone by the qubit's observables. We have further assumed that the modes of the waveguide constitute a large reservoir, so that we could approximate $\sqrt{\rho(\omega)}g(\omega)$ by $\sqrt{\gamma/2\pi}$ [22].

The mean value $b_\mathrm{in}(t)$ of the input operator is expressed in units of the square root of a photon rate. Defining the dimensionless mode

$$\hat{B}_\mathrm{in}(t) := \frac{\hat{b}_\mathrm{in}(t)}{\sqrt{\gamma}} \tag{6.45}$$

allows introducing the input field state

$$|\beta_{\text{in}}(t)\rangle := \mathcal{D}_{\hat{B}_{\text{in}}(t)}(\beta_{\text{in}})|0\rangle, \tag{6.46}$$

where

$$\mathcal{D}_{\hat{a}}(\alpha) := \exp(\alpha^* \hat{a} - \alpha \hat{a}^\dagger) \tag{6.47}$$

is the displacement operator in the mode $\hat{a}$ by the amount $\alpha \in \mathbb{C}$ and $|0\rangle$ is the vacuum state. The number of injected photons reads $N_{\text{in}}(t) = \left\langle \hat{B}_{\text{in}}^\dagger(t)\hat{B}_{\text{in}}(t)\right\rangle$, while the rate of photons impinging on the qubit is $\left\langle \hat{b}_{\text{in}}^\dagger(t)\hat{b}_{\text{in}}(t)\right\rangle = \gamma N_{\text{in}}(t)$. The input drive is a coherent field, resonant with the qubit's frequency and we choose the phase of the input drive so that

$$b_{\text{in}}(t) = |b_{\text{in}}(t)|\mathrm{e}^{-\mathrm{i}\omega_0 t}. \tag{6.48}$$

We have left a dependence in time in $|b_{\text{in}}(t)|$ because in the last part of this chapter we will study the impact of pulse shaping. We obtain the equations of evolution of the population of the excited state $P_e(t)$ and the coherence $s(t)$, defined as the expectation value of the operator σ_-, by replacing O_{q} by $|e\rangle\langle e|$ and σ_- respectively in Eq. (6.42) and taking the average:

$$\dot{P}_e(t) = -\gamma P_e(t) - \Omega(t)\mathrm{Re}(s(t)\mathrm{e}^{\mathrm{i}\omega_0 t}), \tag{6.49a}$$

$$\dot{s}(t) = -\left(\mathrm{i}\omega_0 + \frac{\gamma}{2}\right)s(t) + \Omega(t)\mathrm{e}^{-\mathrm{i}\omega_0 t}\left(P_e(t) - \frac{1}{2}\right), \tag{6.49b}$$

The Rabi frequency is defined as

$$\Omega(t) := 2\sqrt{\gamma}|b_{\text{in}}(t)|. \tag{6.50}$$

6.1.2.2 Input and Output Relations

The output operator $\hat{b}_{\text{out}}(t)$, describing the field in the waveguide after its interaction with the qubit, is related to the input operator by the so-called input-output equation [21]

$$\hat{b}_{\text{out}}(t) = \hat{b}_{\text{in}}(t) + \sqrt{\gamma}\sigma_-(t). \tag{6.51}$$

The operator accounting for the rate of propagating photons in the output field is $\hat{b}_{\text{out}}^\dagger(t)\hat{b}_{\text{out}}(t)$. Using Eq. (6.51), it can be expressed as

$$\hat{b}_{\text{out}}^\dagger(t)\hat{b}_{\text{out}}(t) = \hat{b}_{\text{in}}^\dagger(t)\hat{b}_{\text{in}}(t) + \gamma\sigma_+(t)\sigma_-(t) + \sqrt{\gamma}\left(\hat{b}_{\text{in}}^\dagger(t)\sigma_-(t) + \sigma_+(t)\hat{b}_{\text{in}}(t)\right), \tag{6.52}$$

yielding

$$\left\langle \hat{b}^{\dagger}_{\text{out}}(t)\hat{b}_{\text{out}}(t)\right\rangle = \gamma N_{\text{in}}(t) + \gamma P_e(t) + 2\sqrt{\gamma}\text{Re}(b_{\text{in}}(t)s(t)). \tag{6.53}$$

The input and output powers are defined as

$$P_{\text{in/out}}(t) := \hbar\omega_0 \left\langle \hat{b}^{\dagger}_{\text{in/out}}(t)\hat{b}_{\text{in/out}}(t)\right\rangle. \tag{6.54}$$

From the system of equations (6.49), we can determine that the effective Hamiltonian describing the evolution of the qubit is

$$H_{\text{q}}(t) = \hbar\omega_0|e\rangle\langle e| + \text{i}\frac{\hbar\Omega(t)}{2}(\sigma_- e^{\text{i}\omega_0 t} - \sigma_+ e^{-\text{i}\omega_0 t}). \tag{6.55}$$

Using Eq. (2.21), the mean energy of the qubit is given by

$$\varepsilon(t) = \hbar\omega_0 P_e(t) - \hbar\Omega(t)\text{Im}(s(t)\text{e}^{\text{i}\omega_0 t}). \tag{6.56}$$

By integrating the imaginary part of Eq. (6.49b), we obtain $\text{Im}(s(t)\text{e}^{\text{i}\omega_0 t}) = \text{Im}(s(0))\text{e}^{-\gamma t/2}$. Since the initial state ρ_∞ is such that $s(0)$ is real (See Eq. (6.32)), we have $\text{Im}(s(t)\text{e}^{\text{i}\omega_0 t}) = 0$ at any time t and, therefore

$$\varepsilon(t) = \hbar\omega_0 P_e(t). \tag{6.57}$$

Then, using Eqs. (6.51) and (6.49a), we get the following input-output relations

$$b_{\text{out}}(t) = b_{\text{in}}(t) + \sqrt{\gamma}s(t), \tag{6.58}$$

$$P_{\text{out}}(t) = P_{\text{in}}(t) - \dot{\varepsilon}(t), \tag{6.59}$$

with $b_{\text{out}}(t) = \left\langle \hat{b}_{\text{out}}(t)\right\rangle$. As a consequence, the power emitted by the qubit can be directly accessed by measuring the difference between the output and input powers.

From now on, except in the last part of this chapter, we assume that the rate of incoming photons γN_{in} is constant and, therefore, so is the Rabi frequency $\Omega = 2\gamma\sqrt{N_{\text{in}}}$. In this case, we recover the usual Hamiltonian for a driven qubit and the system of equations (6.49) giving the evolution of the population and coherence become the usual Bloch equations [20].

6.1.3 Two-Stroke Engine

We now combine the two evolutions described above to create a two-stroke engine (See Fig. 6.1b and c):

① The qubit is put in contact with the engineered bath that makes it relax in state ρ_∞ (Eq. (6.32)).

(2) The bath is disconnected and the qubit is coherently driven during a time τ, ending in the state $\rho(\tau)$, obtained from the equations of evolution (6.49).

6.1.3.1 First Stroke

During the first stroke, the qubit receives an energy $\mathcal{E}_{\text{in}}$ from the bath. This energy plays a similar role to heat since it is exchanged during a non-unitary process. The stroke is long enough so that the qubit reaches the steady state ρ_∞ that contains the energy

$$\varepsilon_\infty := \text{Tr}(\rho_\infty H_0) = \frac{\hbar\omega_0}{2}(1 + (2p-1)\cos(\theta)), \tag{6.60}$$

and, unusually, also contains the coherence

$$s_\infty := \text{Tr}(\rho_\infty \sigma_-) = \left(\frac{1}{2} - p\right)\sin(\theta), \tag{6.61}$$

in the qubit's energy eigenbasis $\{|e\rangle, |g\rangle\}$. Therefore part of ε_∞ can be directly extracted by unitary processes, i.e. in the form of work, which is a major difference with respect to thermal baths. The maximal amount of such extractable work, called ergotropy [23], equals in the present case

$$\mathcal{W}_\infty = \hbar\omega_0(1-2p)\sin^2\left(\frac{\theta}{2}\right). \tag{6.62}$$

The ratio $\mathcal{W}_\infty/\varepsilon_\infty$ is represented in Fig. 6.3a in the Bloch representation. This figure shows that the thermal states, corresponding to $\langle\sigma_z\rangle \in [-1, 0]$, $\langle\sigma_x\rangle = \langle\sigma_y\rangle = 0$ in the Bloch sphere, contain no ergotropy, while reciprocally the energy contained

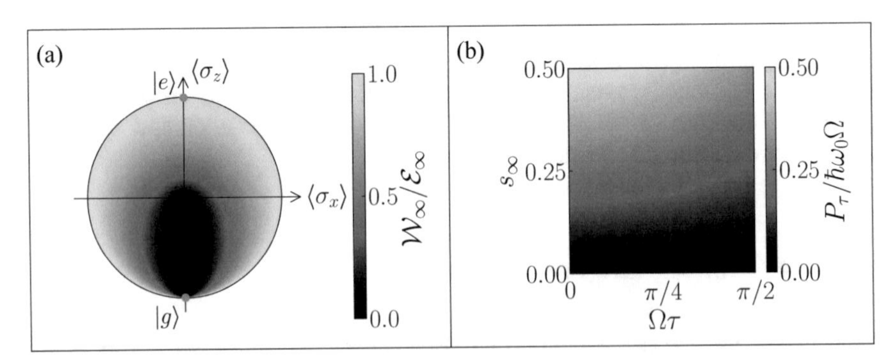

Fig. 6.3 **a** Ergotropy divided by energy as a function of the state ρ_∞ in the Bloch representation. **b** Mean extracted power as a function of the cycle's duration and coherence s_∞ of the state ρ_∞ for $\theta = \pi/2$. Both plots correspond to the stimulated regime ($N_{\text{in}} \gg 1$) where all the energy provided by the bath is converted into work

by a pure state can be fully extracted as work. The typical protocol to fully extract this ergotropy would be to perform a classical Rabi oscillation to bring the qubit back in a thermal state. Indeed, such an evolution is unitary, so, from a thermodynamic point of view, the qubit's energy changes while its entropy remains constant, which characterizes the exchange of work. Since the engineered bath prepares the qubit in a non-passive state, i.e. a state with a non zero ergotropy, it is possible to extract work cyclically from this single non-thermal bath.

6.1.3.2 Second Stroke

The second stroke is the one during which the work is extracted. The qubit evolves during a time τ according to Eqs. (6.49a) and (6.49b). In the stimulated regime $\Omega \gg \gamma$ and in the limit $t \ll \gamma^{-1}$, these equations are reduced to a classical Rabi oscillation. Therefore, the qubit's dipole does not fluctuate and the emitted field reduces to its coherent component, that coherently adds to the input drive. That is why we have treated the electromagnetic mode propagating in the waveguide as a resonant battery. It stores the work released by the working substance in the form of a coherent state, the stimulated regime being equivalent to a large initial load, $N_{\text{in}} \gg 1$. So far, most quantum engines were studied in this regime [6, 24–26].

Interestingly, the one-dimensional geometry also allows driving the qubit in the regime of small N_{in}, i.e. tuning the "load" of the battery to eventually bring it in the quantum regime. The system of equations (6.49) now involves non-unitary energy exchanges. The output power, given by (6.59) can be split into a coherent part and an incoherent part [20, 26]:

$$P_{\text{out}}(t) = \hbar\omega_0 \left(|b_{\text{out}}(t)|^2 + \left\langle \delta\hat{b}^\dagger_{\text{out}}(t)\delta\hat{b}_{\text{out}}(t) \right\rangle \right), \tag{6.63}$$

where $\delta\hat{b}_{\text{out}}$ corresponds to the quantum fluctuations of the output field, such that $\left\langle \delta\hat{b}_{\text{out}} \right\rangle = 0$. By analogy with the stimulated regime, the work rate $\dot{W}$ is identified with the coherent fraction of the emitted power $\dot{\varepsilon}$, while the heat rate $\dot{Q}$ accounts for the power dissipated by the dipole's fluctuations [26]:

$$\begin{aligned} \dot{W}(t) &:= \hbar\omega_0(|b_{\text{out}}(t)|^2 - |b_{\text{in}}(t)|^2) \\ &= \hbar\omega_0 \left(\gamma|s(t)|^2 + \Omega \text{Re}(s(t)e^{i\omega_0 t}) \right), \end{aligned} \tag{6.64}$$

$$\begin{aligned} \dot{Q}(t) &:= \hbar\omega_0 \left\langle \delta\hat{b}^\dagger_{\text{out}}(t)\delta\hat{b}_{\text{out}}(t) \right\rangle \\ &= \hbar\omega_0\gamma(P_e(t) - |s(t)|^2), \end{aligned} \tag{6.65}$$

with $-\dot{\varepsilon}(t) = \dot{W}(t) + \dot{Q}(t)$ and where we have used Eq. (6.58).

The expression of the work rate, Eq. (6.64), shows that part of the spontaneous emission of the qubit is coherent. This spontaneous component scales like γ and can be detected using one-dimensional atoms where the field radiated by the qubit can be collected with high efficiency, and analyzed by using standard homodyning or heterodyning techniques as experimentally demonstrated in Ref. [26]. The coherent fraction of the field in the waveguide provides a new implementation of a resonant quantum battery [15, 27–30]. The battery not only acts as a work repository, but also drives the system, therefore the modeling of the work extraction step does not involve any external operator, like in the previous chapters, though the battery was dispersive.

6.1.3.3 Summary

To summarize the energy exchanges, during the first stroke, an energy $\mathcal{E}_{\text{in}}$ is provided by the bath to the qubit. During the second stroke, $t \in [0, \tau]$, the bath is switched off and the qubit's state evolves following Eqs. (6.49a) and (6.49b), such that $\rho_\infty \to \rho(\tau)$. An amount of work

$$W = \int_0^\tau \mathrm{d}t \hbar\omega_0 \left(\gamma|s(t)|^2 + \Omega \mathrm{Re}(s(t)e^{\mathrm{i}\omega_0 t})\right) \tag{6.66}$$

is extracted in the drive while the heat

$$Q = \int_0^\tau \mathrm{d}t \hbar\omega_0 \gamma (P_e(t) - |s(t)|^2) \tag{6.67}$$

is dissipated in the waveguide. Energy conservation yields

$$\mathcal{E}_{\text{in}} = W + Q. \tag{6.68}$$

By playing on the parameters of the engineered bath, the first stroke can be made arbitrarily short, so that its duration is negligible compared to the other relevant time scales γ^{-1}, Ω^{-1} and τ. Moreover, the input drive used during the second stroke is always on and its action can be neglected during the first stroke provided that the coupling strength between the qubit and the engineered bath is strong enough.

6.2 Classical Battery

We study the engine's performances when the battery is classical, i.e. $N_{\text{in}} \gg 1$, starting with an energetic analysis. Then we analyze this engineered bath powered engine (EBE) as an autonomous version of the measurement powered engine (MPE) proposed in Ref. [25]. We finally focus on the entropy production over one cycle.

6.2.1 Energetic Analysis

In this regime, the work exchanged during one cycle reduces to its stimulated component,

$$W = \hbar\omega_0 \int_0^{\mathcal{T}} \mathrm{d}t \, \Omega \mathrm{Re}(s(t)e^{\mathrm{i}\omega_0 t}). \quad (6.69)$$

The extracted power, $P_{\mathcal{T}} := W/\mathcal{T}$, is plotted in Fig. 6.3b as a function of the cycle's duration $\mathcal{T}$ and input state coherence s_∞ for $\theta = \pi/2$. $P_{\mathcal{T}}$ increases with s_∞ and decreases with $\mathcal{T}$. In the limit of infinitely short cycles, we have $P_{\mathcal{T}} \to P_0 = \Omega\hbar\omega_0 s_\infty$. Naturally, P_0 is maximal when $s_\infty = 1/2$, i.e. for $\rho_\infty = |+_{\pi/2}\rangle\langle+_{\pi/2}|$ since this state gives rise to the maximal slope of the Rabi oscillation (See Fig. 6.1c). This effect is the origin of "coherence induced power boosts" predicted in [5, 25] and reported for an ensemble of qubits in [6]. Using a one-dimensional atom holds the promise of observing such power boosts in the single qubit regime, which has remained elusive so far.

For such classical battery the engine's yield η_{cl} is usually defined by comparing the extracted work to the resource consumed, that is the energy provided by the bath. Thus, $\eta_{\mathrm{cl}} = |W/\mathcal{E}_{\mathrm{in}}|$, which yields here $\eta_{\mathrm{cl}} = 1$ (See Fig. 6.1c): All the energy input by the bath is coherently added to the classical field. Note that η_{cl} does not involve any temperature since the bath is not thermal. The present situation strikingly illustrates that yield and reversibility are independent figures of merit when engines are fueled by non-thermal resources [4]: This engine operates at maximal yield even though it involves an irreversible relaxation step, as detailed in Sect. 6.2.3.

6.2.2 Comparison with a Measurement Powered Engine

We now analyze this device as an autonomous version of the MPE proposed in Ref. [25]. This engine is a four-stroke engine (Fig. 6.4c and d) whereas the EBE is a two-stroke engine (Fig. 6.4a and b) that does not require a state dependent feedback to close the cycle. Nevertheless, both engines are very similar. In order to make a full analogy, we generalized the MPE to the case where the demon's memory, modeled by a two-level system of energy eigenstates $|0\rangle$ and $|1\rangle$, is not prepared in a perfectly pure state but in the thermal mixture

$$\rho_\infty^{\mathcal{D}} = p_{\mathcal{D}}|1\rangle\langle 1| + (1 - p_{\mathcal{D}})|0\rangle\langle 0|. \quad (6.70)$$

$p_{\mathcal{D}}$ is the equilibrium population of the excited state for a temperature $T_{\mathcal{D}}$. The states and operators relating to the qubit (resp. demon) are denoted by the label $\mathcal{S}$ (resp. $\mathcal{D}$).

For the MPE, $\{|-_\theta\rangle, |+_\theta\rangle\}$ is the measurement basis while for the EBE, it is the basis imposed by the engineered bath. The first stroke of the EBE (relaxation

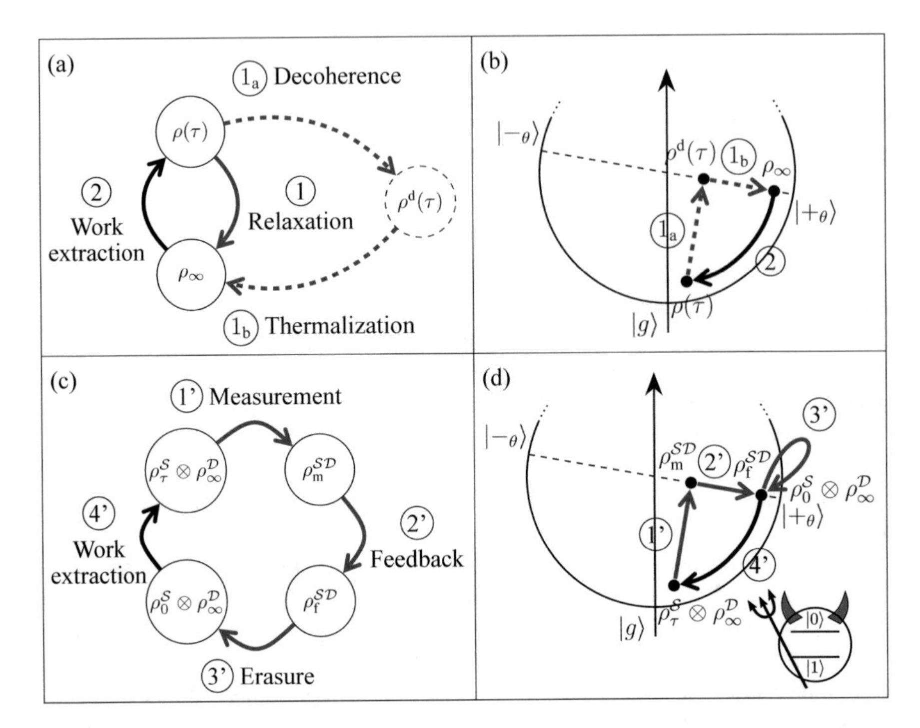

Fig. 6.4 Cycles of **a** the engine proposed in this chapter (EBE) and of **c** the measurement powered engine (MPE) from [25]. See Table 6.1 for the expressions of the density operators. **b** and **d** represent the same cycles as **a** and **c** respectively, but in the Bloch sphere. The dashed arrows in **a** and **b** represent the effective decomposition of the first stroke into a decoherence step and a thermalization step (See Sect. 6.2.3.2)

of the qubit in contact with the engineered bath) is equivalent to the three first strokes (measurement, feedback and erasure) of the MPE. For both engines, the work extraction is performed during the last stroke by driving the qubit during a time τ. The temperature of the demon's memory corresponds to the effective temperature of the engineered bath, which can be defined from the equilibrium populations of the states $|\pm_\theta\rangle$. The limit of infinitely short cycles for the EBE is equivalent to the Zeno regime for the MPE.

The analogy between the two engines, detailed in Table 6.1, is even stronger when decomposing the relaxation step of the EBE into a decoherence step and a thermalization step. This decomposition is motivated by the expression of the entropy production which can be split into a decoherence term and a thermalization term, as shown below.

Table 6.1 Step by step analogy between the EBE and MPE. We have used the notations $|\psi_\pm\rangle = U(\tau)|\pm_\theta\rangle$ where U is the evolution operator of the driven qubit

<table>
<tr><th>Engineered bath powered engine (EBE)</th><th>Measurement powered engine (MPE)</th></tr>
<tr><td>(1a) Decoherence $\rho(\tau) \to \rho_{\mathrm{d}}(\tau)$
The qubit, initially in state
$\rho(\tau) = p|\psi_-\rangle\langle\psi_-| + (1-p)|\psi_+\rangle\langle\psi_+|$,
loses its coherences in the $\{|+_\theta\rangle, |-_\theta\rangle\}$ basis:
$\rho_{\mathrm{d}}(\tau) = (1-q)|+_\theta\rangle\langle+_\theta| + q|-_\theta\rangle\langle-_\theta|$
$q = \langle\rho(\tau)\rangle_{-\theta}$ and $1-q = \langle\rho(\tau)\rangle_{+\theta}$</td>
<td>(1') Measurement $\rho_\tau^{\mathcal{S}} \otimes \rho_\infty^{\mathcal{D}} \to \rho_{\mathrm{m}}^{\mathcal{SD}}$
The qubit, initially in state
$\rho_\tau^{\mathcal{S}} = p_{\mathcal{D}}|\psi_-\rangle\langle\psi_-| + (1-p_{\mathcal{D}})|\psi_+\rangle\langle\psi_+|$,
is measured in the $\{|+_\theta\rangle, |-_\theta\rangle\}$ basis and correlated to the demon's memory via a controlled-NOT operation:
$\rho_{\mathrm{m}}^{\mathcal{SD}} = (1-p_{\mathcal{D}})(1-q)|+_\theta, 1\rangle\langle+_\theta, 1|$
$+p_{\mathcal{D}}q|-_\theta, 1\rangle\langle-_\theta, 1|$
$+p_{\mathcal{D}}(1-q)|+_\theta, 0\rangle\langle+_\theta, 0|$
$+(1-p_{\mathcal{D}})q|-_\theta, 0\rangle\langle-_\theta, 0|$
$q = \langle\rho_\tau^{\mathcal{S}}\rangle_{-\theta}$ and $1-q = \langle\rho_\tau^{\mathcal{S}}\rangle_{+\theta}$</td></tr>
<tr><td>(1b) Thermalization $\rho_{\mathrm{d}}(\tau) \to \rho_\infty$
The populations of states $|\pm_\theta\rangle\langle\pm_\theta|$ go to their equilibrium values</td>
<td>(2') Feedback $\rho_{\mathrm{m}}^{\mathcal{SD}} \to \rho_{\mathrm{f}}^{\mathcal{SD}}$
The transformation $\mathbf{1}^{\mathcal{S}} \otimes |1\rangle\langle 1| + U_\pi^{\mathcal{S}} \otimes |0\rangle\langle 0|$ is applied, where $\mathbf{1}^{\mathcal{S}}$ is the identity and $U_\pi^{\mathcal{S}}$ is a π-pulse. Therefore,
$\rho_{\mathrm{f}}^{\mathcal{SD}} = (1-p_{\mathcal{D}})(1-q)|+_\theta, 1\rangle\langle+_\theta, 1|$
$+p_{\mathcal{D}}q|-_\theta, 1\rangle\langle-_\theta, 1|$
$+p_{\mathcal{D}}(1-q)|-_\theta, 0\rangle\langle-_\theta, 0|$
$+(1-p_{\mathcal{D}})q|+_\theta, 0\rangle\langle+_\theta, 0|$
and $\mathrm{Tr}_{\mathcal{D}}(\rho_{\mathrm{f}}^{\mathcal{SD}}) = \rho_0^{\mathcal{S}}$ where
$\rho_0^{\mathcal{S}} = p_{\mathcal{D}}|-_\theta\rangle\langle-_\theta| + (1-p_{\mathcal{D}})|+_\theta\rangle\langle+_\theta|$.
(3') Erasure of the memory
$\rho_{\mathrm{f}}^{\mathcal{SD}} \to \rho_0^{\mathcal{S}} \otimes \rho_\infty^{\mathcal{D}}$
The demon's memory is put in contact with a bath at temperature $T_{\mathcal{D}}$ for long enough to thermalize</td></tr>
<tr><td>(2) Work extraction $\rho_\infty \to \rho(\tau)$</td>
<td>(4') Work extraction $\rho_0^{\mathcal{S}} \otimes \rho_\infty^{\mathcal{D}} \to \rho_\tau^{\mathcal{S}} \otimes \rho_\infty^{\mathcal{D}}$</td></tr>
</table>

6.2.3 *Entropic Analysis*

In this part, we study the entropy produced during the first stroke. We generalize the expression of the entropy production in the thermal case, given by Eq. (2.38), to the case where the qubit relaxes in contact with the engineered bath

$$S_{\mathrm{irr}} = S(\rho(\tau)\|\rho_\infty), \tag{6.71}$$

where $S(\rho(\tau)\|\rho_\infty)$ is the relative entropy, defined by Eq. (2.37). In the thermal case, the entropy production can be split into a decoherence part and a thermalization part. Based on this decomposition, we will first show that a thermal relaxation is equivalent to the concatenation of a decoherence map and a thermalization map. Then, by analogy with the thermal case, we will use the same decomposition to compute the entropy production during the relaxation of the qubit in contact with the engineered bath.

6.2.3.1 Effective Map for Thermal Relaxation

In this part, we consider a qubit coupled to a thermal bath at finite temperature. The qubit is initially prepared in the state

$$\rho_0 = p_0|\psi\rangle\langle\psi| + (1 - p_0)|\bar{\psi}\rangle\langle\bar{\psi}|, \tag{6.72}$$

where

$$|\psi\rangle = \alpha|e\rangle + \beta|g\rangle, \tag{6.73}$$

$$|\bar{\psi}\rangle = -\beta^*|e\rangle + \alpha^*|g\rangle. \tag{6.74}$$

α and β are two complex numbers such that $|\alpha|^2 + |\beta|^2 = 1$ and $p_0 \in [0, 1]$. The qubit is put in contact with the thermal bath and relaxes toward the equilibrium state

$$\rho_{\text{eq}} = p_{\text{eq}}|e\rangle\langle e| + (1 - p_{\text{eq}})|g\rangle\langle g|. \tag{6.75}$$

We assume that the duration t_f of the transformation is long enough ($t_\text{f} \gg \gamma^{-1}$) for the qubit to reach the equilibrium state. The average entropy production during this relaxation is thus given by Eq. (2.38),

$$S_{\text{irr}} = S(\rho_0\|\rho_{\text{eq}}), \tag{6.76}$$

which can be split in a decoherence part and a thermalization part

$$S_{\text{irr}} = S_{\text{irr}}^{\text{d}} + S_{\text{irr}}^{\text{th}}. \tag{6.77}$$

These two parts respectively read $S_{\text{irr}}^{\text{d}} = S(\rho_0\|\rho_\text{d})$ and $S_{\text{irr}}^{\text{th}} = S(\rho_\text{d}\|\rho_{\text{eq}})$, where

$$\rho_\text{d} = \langle e|\rho_0|e\rangle\, |e\rangle\langle e| + \langle g|\rho_0|g\rangle\, |g\rangle\langle g|. \tag{6.78}$$

We want to show that this relaxation can be effectively described by the concatenation of a decoherence map $\mathcal{L}^\text{d}$ and a thermalization map $\mathcal{L}^{\text{th}}$. The decoherence and thermalization maps are respectively defined by

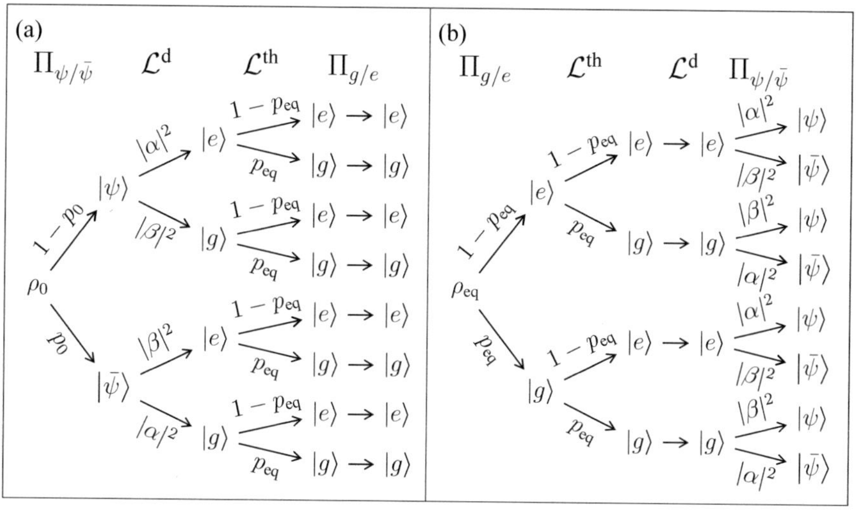

Fig. 6.5 All possible trajectories for **a** the direct protocol and **b** the reversed protocol when modeling the thermal relaxation as the concatenation of a decoherence map and a thermalization map

$$\mathcal{L}^{\mathrm{d}}[\rho] = |g\rangle\langle g|\rho|g\rangle\langle g| + |e\rangle\langle e|\rho|e\rangle\langle e|, \tag{6.79}$$

$$\mathcal{L}^{\mathrm{th}}[\rho] = \sum_{i,j=e,g} M_{ij}\rho M_{ij}^{\dagger}, \tag{6.80}$$

with

$$M_{ij} = \sqrt{\langle i|\rho_{\mathrm{eq}}|i\rangle}|i\rangle\langle j|. \tag{6.81}$$

We will use the definition of entropy production at the single trajectory level, $s_{\mathrm{irr}}[\vec{\Sigma}]$, given by Eq. (2.50), to prove that $\mathcal{L}^{\mathrm{th}} \circ \mathcal{L}^{\mathrm{d}}$ gives the same average entropy production S_{irr}. $P[\vec{\Sigma}]$ (resp. $\tilde{P}[\overleftarrow{\Sigma}]$) is the probability that the qubit follows the trajectory $\vec{\Sigma}$ (resp. $\overleftarrow{\Sigma}$) during the direct (resp. reversed) protocol. The trajectories $\vec{\Sigma} = (\Sigma_0, \Sigma_1, \Sigma_2)$, where $\Sigma_0 \in \{|\psi\rangle, |\bar{\psi}\rangle\}$ and $\Sigma_1, \Sigma_2 \in \{|g\rangle, |e\rangle\}$, are obtained by performing a quantum jump unraveling of the map concatenation. The direct protocol consists in projecting the initial state ρ_0 in the $\{|\psi\rangle, |\bar{\psi}\rangle\}$ basis, then applying the maps $\mathcal{L}^{\mathrm{d}}$ and $\mathcal{L}^{\mathrm{th}}$ successively and finally projecting in the $\{|g\rangle, |e\rangle\}$ basis. The reverse protocol consists in applying the same operations in the reverse order with the qubit initially in ρ_{eq}. Reading the probabilities $P[\vec{\Sigma}]$ (resp. $\tilde{P}[\overleftarrow{\Sigma}]$) of the direct (resp. reversed) trajectories in Fig. 6.5a (resp. Fig. 6.5b), we obtain the average entropy production:

$$\begin{aligned}\left\langle s_{\mathrm{irr}}[\vec{\Sigma}]\right\rangle_{\vec{\Sigma}} &= p_0 \log p_0 + (1-p_0)\log(1-p_0) \\ &\quad - (p_0|\alpha|^2 + (1-p_0)|\beta|^2)\log p_{\mathrm{eq}} - ((1-p_0)|\alpha|^2 + p_0|\beta|^2)\log(1-p_{\mathrm{eq}}) \\ &= S_{\mathrm{irr}}.\end{aligned} \tag{6.82}$$

Therefore this effective decomposition of the relaxation can be used to compute the average entropy production.

6.2.3.2 Entropy Production over One Engine's Cycle

During the first stroke of the engine's cycle, the qubit is put in contact with an engineered bath. Since $\{|-_\theta\rangle, |+_\theta\rangle\}$ is the energy eigenbasis of the qubit in contact with the engineered bath (See Sect. 6.1.1.1), by analogy with the thermal bath case discussed above, the transformation can be split into a decoherence step and a thermalization step (See Fig. 6.4a and b), but in the basis $\{|-_\theta\rangle, |+_\theta\rangle\}$ instead of $\{|e\rangle, |g\rangle\}$. These steps correspond respectively to the maps $\mathcal{L}^{\mathrm{d}}$ and $\mathcal{L}^{\mathrm{th}}$, defined by

$$\mathcal{L}^{\mathrm{d}}[\rho] := |+_\theta\rangle\langle+_\theta|\rho|+_\theta\rangle\langle+_\theta| + |-_\theta\rangle\langle-_\theta|\rho|-_\theta\rangle\langle-_\theta|, \tag{6.83}$$

$$\mathcal{L}^{\mathrm{th}}[\rho] := \sum_{i,j=\pm_\theta} M_{ij}\rho M_{ij}^\dagger, \tag{6.84}$$

with

$$M_{ij} = \sqrt{\langle i|\rho_\infty|i\rangle}|i\rangle\langle j|. \tag{6.85}$$

The probabilities $P[\vec{\Sigma}]$ and $\tilde{P}[\overleftarrow{\Sigma}]$ of the direct and reversed trajectories can be read from diagrams similar to Fig. 6.5a and b, but replacing ρ_0 by $\rho(\tau)$ and ρ_{eq} by ρ_∞, therefore p_0 and p_{eq} are replaced by p. Then, the average entropy production reads

$$\left\langle s_{\mathrm{irr}}[\vec{\Sigma}]\right\rangle_{\vec{\Sigma}} = (1-2p)\sin^2\left(\frac{\Omega\tau}{2}\right)\log\left(\frac{1-p}{p}\right), \tag{6.86}$$

which can be rewritten as a relative entropy, giving Eq. (6.71).

Since we consider the case where $\gamma^{-1} \gg \tau$, the second stroke is reversible and the entropy production over one cycle is given by Eq. (6.71). Like previously, we have

$$S_{\mathrm{irr}} = S_{\mathrm{irr}}^{\mathrm{d}} + S_{\mathrm{irr}}^{\mathrm{th}}, \tag{6.87}$$

where $S_{\mathrm{irr}}^{\mathrm{d}} = S(\rho(\tau)\|\rho_{\mathrm{d}}(\tau))$ is the decoherence contribution, with

$$\rho_{\mathrm{d}}(\tau) = \langle+_\theta|\rho(\tau)|+_\theta\rangle|+_\theta\rangle\langle+_\theta| + \langle-_\theta|\rho(\tau)|-_\theta\rangle|-_\theta\rangle\langle-_\theta|, \tag{6.88}$$

and $S_{\mathrm{irr}}^{\mathrm{th}} = S(\rho_{\mathrm{d}}(\tau)\|\rho_\infty)$ is the thermalization contribution. The entropy production is plotted as a function of the cycle duration τ for different values of p in Fig. 6.6a and the decoherence and thermal contributions are respectively plotted in Fig. 6.6b and c. As expected, there is no entropy production when $p = 0.5$ because ρ_∞ is in the center of the Bloch sphere and therefore the qubit's state does not evolve during the second stroke. On the contrary, the thermal contribution, and thus the total entropy production, diverge when p goes to 0 with $\tau > 0$ because the qubit is in a pure state.

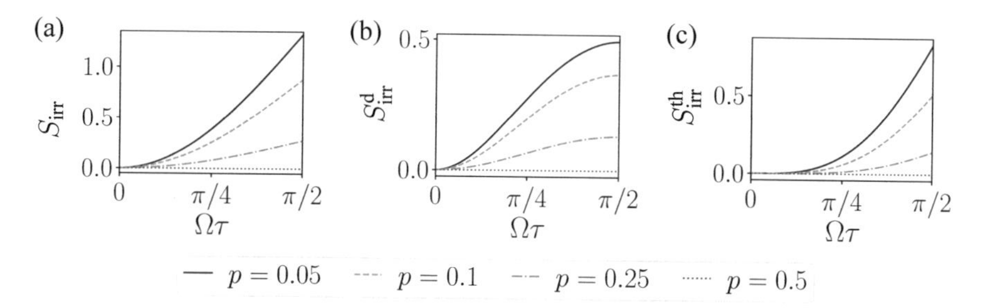

Fig. 6.6 Entropy production: **a** over one engine's cycle, **b** decoherence contribution and **c** thermalization contribution. The entropy production is plotted as a function of the duration τ of the cycle for different values of p

This behavior is analogous to what would be obtained for a thermal relaxation in a thermal bath at zero temperature.

The entropic study presented above invites to draw an analogy between the energy $\mathcal{E}_{\text{in}}$ provided by the bath and the quantum heat introduced in Ref. [25], defined as the energy fluctuations induced by the measurement channel. It can indeed be shown that $\mathcal{E}_{\text{in}}$ is provided to the qubit during the decoherence step, i.e. the measurement performed by the bath on the system. The complete thermodynamic analysis of the energy exchanges between the qubit and the engineered bath is beyond the scope of this chapter, which focuses on the impact of coherence on work extraction, and will be treated in a different project [31].

6.3 Quantum Battery

We now study the engine for arbitrary driving strengths. For the sake of clarity we assume that the bath prepares pure states $\rho_\infty = |+_\theta\rangle\langle+_\theta|$.

6.3.1 *Work and Efficiency*

The extracted work is given by Eq. (6.64) and is plotted in Fig. 6.7a as a function of Ω/γ and θ. At each point we have chosen the duration τ of the cycle that maximizes W. For fixed θ, W increases with Ω/γ because stimulated emission allows for the funneling of energy in the coherent driving mode. Maximal work extraction is reached in the classical limit $\Omega \gg \gamma$, when $\theta = \pi$, i.e. full population inversion. This situation is typical of single qubit lasers and masers [24]. Since stimulated emission has a favorable impact on work extraction, we will now include the energetic cost of loading the battery into the resources used to run the engine, bringing a modified expression of the yield

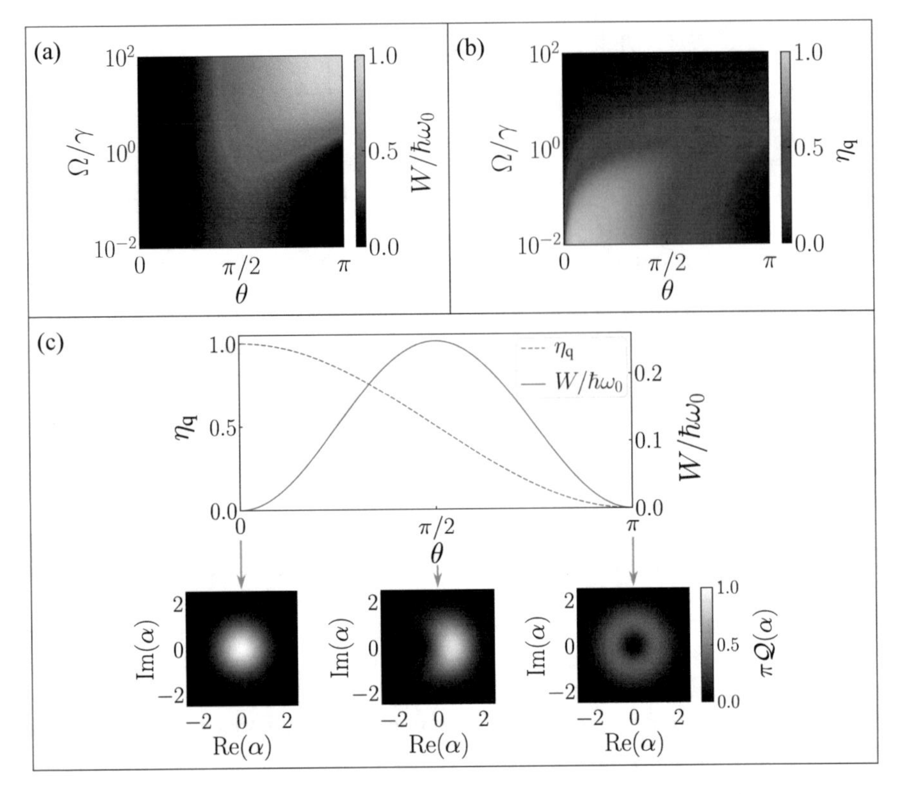

Fig. 6.7 **a** Work and **b** efficiency as functions of the driving strength and the angle θ. The parameters we used are $\gamma = 1$ GHz, $p = 0$ and the cycle time τ maximizing the amount of work extracted for each value of Ω and θ. **c** Extracted work and efficiency in the spontaneous regime ($\Omega = 0$) as functions of the angle θ

$$\eta_q := \frac{|W|}{\mathcal{E}_{in} + \int_0^\tau dt\, P_{in}(t)}. \tag{6.89}$$

η_q is plotted in Fig. 6.7b. It vanishes in the limit $\Omega \gg \gamma$, evidencing that the classical regime is not thermodynamically efficient when the battery's preparation is considered in the balance. Conversely the case where $\Omega \leq \gamma$ gives rise to the largest yields and non-negligible work extraction, but requires some coherence to be initially injected in the qubit: Quantum coherence thus acts as a genuinely quantum resource, that plays a similar role as stimulated emission.

6.3.2 Spontaneous Regime

To single out the effect of quantum coherence on work extraction, we focus on the spontaneous regime ($\Omega = 0$) in the limit $\tau \gg \gamma^{-1}$. The bath prepares $\rho_\infty = |+_\theta\rangle\langle+_\theta|$, providing the energy ε_∞ to the qubit. A fraction of ε_∞ is then released as spontaneous work. In the end of the cycle the qubit has relaxed in $|g\rangle$. Integrating Eq. (6.64) with $\Omega = 0$ yields

$$W = \hbar\omega_0 s_\infty^2, \tag{6.90}$$

revealing a fundamental and so far overlooked relation between work and coherence. Conversely, the engine's yield reads

$$\eta_\mathrm{q} = \hbar\omega_0 \frac{s_\infty^2}{\varepsilon_\infty}. \tag{6.91}$$

Both quantities are plotted in Fig. 6.7c as a function of θ. They decrease with $\theta \geq \pi/2$, and vanish when $\theta = \pi$. Conversely for $\theta \leq \pi/2$, W and η_q cannot be optimized simultaneously.

These behaviors acquire an intuitive interpretation by invoking the nature of the quantum state of light spontaneously emitted in the waveguide during the process. It reads

$$|\psi_\mathrm{out}(\theta)\rangle = \cos\left(\frac{\theta}{2}\right)|0\rangle + \sin\left(\frac{\theta}{2}\right)|1\rangle, \tag{6.92}$$

where

$$|n\rangle := \frac{(\hat{b}^\dagger)^n}{\sqrt{n!}}|0\rangle \tag{6.93}$$

is the n-photon Fock state in the mode defined as

$$\hat{b} := \sqrt{\gamma}\int_0^\tau \mathrm{d}t\, \hat{b}_\mathrm{out}(t). \tag{6.94}$$

The state $|\psi_\mathrm{out}(\theta)\rangle$ partially overlaps with the coherent field $|\beta_\theta\rangle$ of amplitude

$$\beta_\theta = \langle\psi_\mathrm{out}(\theta)|\hat{b}|\psi_\mathrm{out}(\theta)\rangle. \tag{6.95}$$

This coherent field carries the energy

$$\hbar\omega_0|\beta_\theta|^2 = \hbar\omega_0\frac{\sin^2(\theta)}{4}, \tag{6.96}$$

which corresponds to the extracted work. Conversely, the yield quantifies the overlap between $|\beta_\theta\rangle$ and the emitted quantum field $|\psi_\mathrm{out}(\theta)\rangle$. It reads

$$\eta_q = \frac{1 + \cos(\theta)}{2}. \tag{6.97}$$

We have plotted as insets of Fig. 6.7c the Husimi $\mathcal{Q}$ function of $|\psi_{\rm out}(\theta)\rangle$ for $\theta = \{0, \pi/2, \pi\}$. It is defined as [32]

$$\mathcal{Q}_\theta(\alpha) = \frac{1}{\pi} |\langle \psi_{\rm out}(\theta)|\alpha\rangle|^2 \tag{6.98}$$

and characterizes its overlap with a coherent field $|\alpha\rangle$. $\mathcal{Q}_\theta(\beta_\theta)$ and η_q vanish for $\theta = \pi$ where $\beta_\theta = 0$. This is consistent with the fact that single photons have no phase. Therefore a single photon source gives rise to no work extraction. Conversely $\mathcal{Q}_\theta(\beta_\theta)$ and η_q reach 1 when θ goes to 0. This characterizes that $|\psi_{\rm out}(\theta)\rangle$ is fully coherent, however the work extracted vanishes in this limit. The case $\theta = \pi/2$, where the coherence is maximal, offers an interesting trade-off since it maximizes the work extraction $W = \hbar\omega_0/4$, keeping a finite value of the yield $\eta_q = 1/2$.

6.3.3 *Pulse Shape Optimization*

The above study explains why quantum coherence and stimulated emission both contribute to work extraction. In the stimulated regime, the classical phase of the coherent field partially radiated by the dipole is fixed by the classical phase of the drive. In the spontaneous regime, it is fixed by the quantum phase of the initial qubit's state, which requires the initial injection of coherence. To investigate further the interplay between the load of the battery and the coherence injected by the bath, we now consider the following scenario: At $t = 0$ the qubit is prepared in the state ρ_∞ with $\theta = \pi/2$. It is then coupled to a coherent pulsed field giving rise to the Rabi frequency $\Omega(t)$ (Eq. (6.50)). The pulse contains a fixed mean number of photons

$$N_{\rm in} = \frac{1}{4\gamma} \int_0^\infty {\rm d}t\, \Omega^2(t). \tag{6.99}$$

Work and heat are collected until the qubit has fully relaxed, i.e. during a time $\tau \gg \gamma^{-1}$.

The extracted work W and the yield

$$\eta_q = \frac{|W|}{\varepsilon_\infty + \hbar\omega_0 N_{\rm in}} \tag{6.100}$$

are plotted in Fig. 6.8a and b respectively as functions of $N_{\rm in}$ and s_∞ in the case of a rectangular pulse of duration γ^{-1}, for $N_{\rm in} \leq 1$. Both plots evidence that the energy initially contained in the battery and the quantum coherence injected in the qubit act as complementary resources potentially enhancing the engine's performances. While maximal coherence maximizes both figures of merit, the work extraction

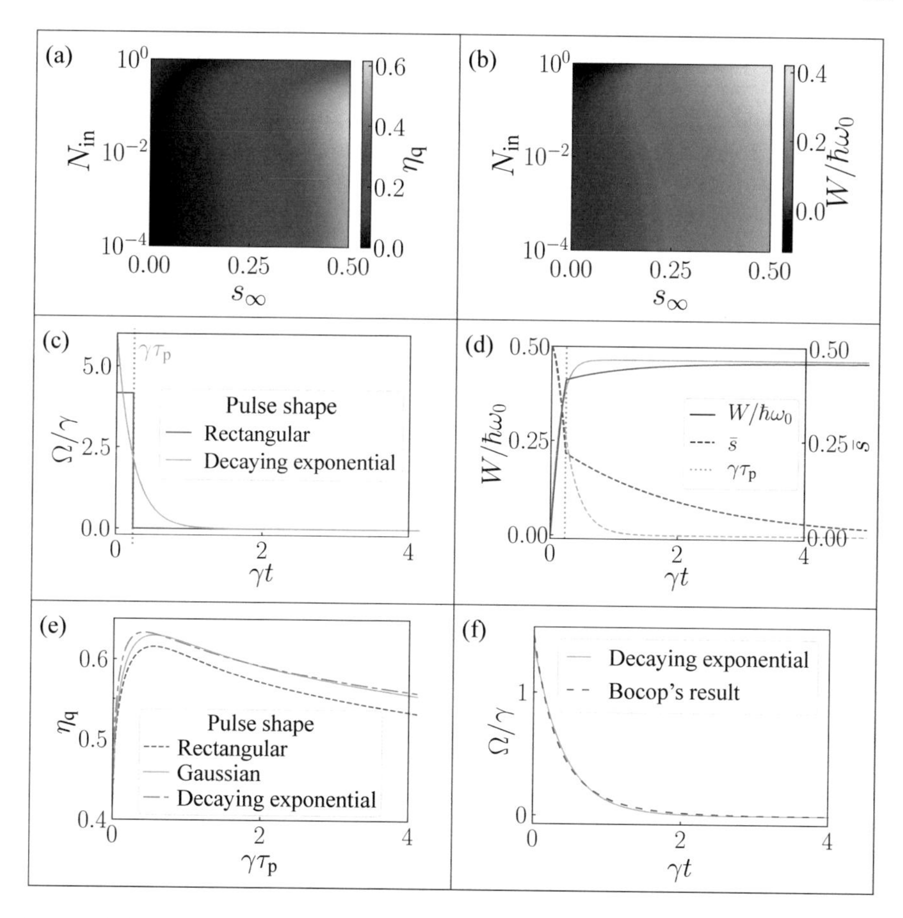

Fig. 6.8 Influence of the pulse shape. **a** Yield and **b** work extracted as a function of $N_{\rm in}$ and s_∞ for $\theta = \pi/2$ and a pulsed square drive of duration $\tau_{\rm p} = \gamma^{-1}$. **c** Pulse shapes and **d** corresponding extracted work (solid lines) and coherence $\bar{s}(t) = s(t)e^{i\omega_0 t}$ (dashed lines) as functions of time. The color of the curves in **d** indicate the pulse shape: rectangular in blue and decaying exponential in orange. *Parameters:* $N_{\rm in} = 1$ and the $\tau_{\rm p}$ maximizing work extraction for the decaying exponential ($\gamma\tau_{\rm p} \simeq 0.17$). **e** Yield as a function of the duration of the pulse for different pulse shapes, with $N_{\rm in} = 0.1$. **f** Comparison between the optimal pulse shape numerically computed by the software Bocop (solid orange) and the exponentially decaying pulse of optimal characteristic time $\tau_{\rm p} \simeq 0.41/\gamma$ (dashed blue). *Parameters:* $N_{\rm in} = 0.1$ and $p = 0$. We took $\gamma = 1$ GHz for all plots

(resp. the yield) is maximal with $N_{\rm in} \sim 1$ (resp. $N_{\rm in} \sim 1/10$). We have then fixed $N_{\rm in} = 1/10$ and studied the impact of the shape of the pulse on the performance. A rectangular pulse and an exponentially decaying pulse of identical characteristic time $\tau_{\rm p} \simeq 0.17/\gamma$ are plotted in Fig. 6.8c. This value of $\tau_{\rm p}$ is the one maximizing the amount of work extracted with an exponentially decaying pulse. The corresponding temporal evolution of the work and coherence are represented in Fig. 6.8d. These plots evidence that the exponentially decaying pulse allows a faster and slightly larger work extraction. Furthermore, the exponentially decaying pulse also give rise to a larger yield than rectangular and Gaussian pulses, as shown in Fig. 6.8e.

After studying three specific pulse shapes, we want to find the pulse shape that maximizes the yield η_q, which is equivalent to maximizing the extracted work W since the number of photons is fixed. This is an optimal control problem where the quantity to maximize is

$$W = \int_0^{\mathcal{T}} dt \left(\Omega(t)\mathrm{Re}(s(t)e^{i\omega_0 t}) + \gamma |s(t)|^2\right) \tag{6.101}$$

and the constraints are Eqs. (6.99), (6.49a) and (6.49b). We solved this problem numerically using Bocop [33] and found that the yield is optimized for a decaying exponential of typical duration $\tau_p \leq \gamma^{-1}$, as shown in Fig. 6.8f. This effect was already observed in the context of optimal irreversible stimulated emission [34] and corresponds to the optimal mode matching between the drive and the qubit. The search for mode matching here finds a new application in the field of quantum thermodynamics.

6.4 Summary

In this chapter, we have evidenced that exploiting advanced tools of quantum optics like engineered baths and batteries opens a new regime for the study of quantum engines, where stimulated emission and quantum coherence behave as complementary resources. We have more specifically studied a two-stroke engine extracting work cyclically from a single non-thermal bath.

First, we derived the evolution of the qubit during the first stroke, the relaxation of the qubit in contact with an non-thermal bath. We showed how the bath can be engineered to prepare the qubit in an arbitrary state containing coherence in the energy eigenbasis. Secondly, we focused on the second stroke, and described the evolution of the qubit embedded in a one-dimensional waveguide using the input-output formalism. We evidenced that the modes of the waveguide resonant with the qubit's transition can be used as a resonant battery to drive the qubit. Importantly, the load of this battery which corresponds to the input photon rate, is tunable and our description is valid for any photon number N_{in}, from the classical regime $N_{in} \gg 1$ to the spontaneous regime $N_{in} = 0$. In this framework, work is the coherent fraction of the energy emitted by the qubit in the waveguide, while heat is dissipated through the dipole fluctuations. Work can be directly accessed in the battery by measuring the output field using homodyning or heterodyning techniques.

Then, we studied the stimulated regime $\Omega \gg \gamma$ which also corresponds to the classical regime $N_{in} \gg 1$. We demonstrated that coherence boosts power extraction, which is maximal when the duration of cycle goes to zero. Furthermore, in this regime, the engine always operates at the maximal classical yield even though it involves an irreversible relaxation. We showed that this device can be seen as an autonomous version of the measurement driven engine proposed in Ref. [25].

Finally, we explored arbitrary driving strengths, redefining the yield to take into account the initial load of the battery. We demonstrated that the stimulated regime is thermodynamically inefficient due to the cost of loading the battery. Conversely, the weak driving limit gives rise to the largest yields but requires some initial coherence. In the spontaneous regime, i.e. in the total absence of driving, coherence allows controlling the engine's yield, acting as a genuinely quantum resource. Lastly, we studied the impact of the pulse shape on work extraction. We evidenced that the optimal pulse shape is a decaying exponential, which corresponds to the optimal mode matching between the drive and the qubit.

References

1. Scully MO, Zubairy MS, Agarwal GS, Walther H (2003) Extracting work from a single heat bath via vanishing quantum coherence. Science 299(5608):862–864. https://doi.org/10.1126/science.1078955
2. Roßnagel J, Abah O, Schmidt-Kaler F, Singer K, Lutz E (2014) Nanoscale heat engine beyond the Carnot limit. Phys Rev Lett 112(3):030602. https://doi.org/10.1103/PhysRevLett.112.030602
3. Klaers J, Faelt S, Imamoglu A, Togan E (2017) Squeezed thermal reservoirs as a resource for a nanomechanical engine beyond the Carnot limit. Phys Rev X 7(3):031044. https://doi.org/10.1103/PhysRevX.7.031044
4. Niedenzu W, Mukherjee V, Ghosh A, Kofman AG, Kurizki G (2018) Quantum engine efficiency bound beyond the second law of thermodynamics. Nat Commun 9(1):165. https://doi.org/10.1038/s41467-017-01991-6
5. Uzdin R, Levy A, Kosloff R (2015) Equivalence of quantum heat machines, and quantum-thermodynamic signatures. Phys Rev X 5(3):031044. https://doi.org/10.1103/PhysRevX.5.031044
6. Klatzow J, Becker JN, Ledingham PM, Weinzetl C, Kaczmarek KT, Saunders DJ, Nunn J, Walmsley IA, Uzdin R, Poem E (2019) Experimental demonstration of quantum effects in the operation of microscopic heat engines. Phys Rev Lett 122(11):110601. https://doi.org/10.1103/PhysRevLett.122.110601
7. Hoi I-C, Palomaki T, Lindkvist J, Johansson G, Delsing P, Wilson CM (2012) Generation of nonclassical microwave states using an artificial atom in 1D open space. Phys Rev Lett 108(26):263601. https://doi.org/10.1103/PhysRevLett.108.263601
8. Eichler C, Lang C, Fink JM, Govenius J, Filipp S, Wallraff A (2012) Observation of entanglement between itinerant microwave photons and a superconducting qubit. Phys Rev Lett 109(24):240501. https://doi.org/10.1103/PhysRevLett.109.240501
9. Giesz V, Somaschi N, Hornecker G, Grange T, Reznychenko B, De Santis L, Demory J, Gomez C, Sagnes I, Lemaítre A, Krebs O, Lanzillotti-Kimura ND, Lanco L, Aufféves A, Senellart P (2016) Coherent manipulation of a solid-state artificial atom with few photons. Nat Commun 7:11986. https://doi.org/10.1038/ncomms11986
10. Ding D, Appel MH, Javadi A, Zhou X, Löbl MC, Söllner I, Schott R, Papon C, Pregnolato T, Midolo L, Wieck AD, Ludwig A, Warburton RJ, Schröder T, Lodahl P (2019) Coherent optical control of a quantum-dot spin-qubit in a waveguide-based spin-photon interface. Phys Rev Appl 11(3):031002. https://doi.org/10.1103/PhysRevApplied.11.031002
11. Monsel J, Fellous-Asiani M, Huard B, Aufféves A (2020) The energetic cost of work extraction. Phys Rev Lett 124(13):130601. https://doi.org/10.1103/PhysRevLett.124.130601
12. Poyatos JF, Cirac JI, Zoller P (1996) Quantum reservoir engineering with laser cooled trapped ions. Phys Rev Lett 77(23):4728–4731. https://doi.org/10.1103/PhysRevLett.77.4728

13. Elouard C, Bernardes NK, Carvalho ARR, Santos MF, Aufféves A (2017) Probing quantum fluctuation theorems in engineered reservoirs. New J Phys 19(10):103011. https://doi.org/10.1088/1367-2630/aa7fa2
14. Tacchino F, Aufféves A, Santos MF, Gerace D (2018) Steady state entanglement beyond thermal limits. Phys Rev Lett 120(6):063604. https://doi.org/10.1103/PhysRevLett.120.063604
15. Barra F (2019) Dissipative charging of a quantum battery. Phys Rev Lett 122(21):210601. https://doi.org/10.1103/PhysRevLett.122.210601
16. Murch KW, Vool U, Zhou D, Weber SJ, Girvin SM, Siddiqi I (2012) Cavity-assisted quantum bath engineering. Phys Rev Lett 109(18):183602+. https://doi.org/10.1103/physrevlett.109.183602
17. Harrington PM, Naghiloo M, Tan D, Murch KW (2019) Bath engineering of a fluorescing artificial atom with a photonic crystal. Phys Rev A 99(5):052126. https://doi.org/10.1103/PhysRevA.99.052126
18. Breuer H-P, Petruccione F (2002) The theory of open quantum systems. Oxford University Press, New York
19. Elouard C (2017) Thermodynamics of quantum open systems: applications in quantum optics and optomechanics. PhD thesis, Université Grenoble Alpes
20. Cohen-Tannoudji C, Dupont-Roc J, Grynberg G (2004) Atom-photon interactions: basic processes and applications, ser. Physics textbook. Wiley, Weinheim-VCH
21. Gardiner CW, Collett MJ (1985) Input and output in damped quantum systems: quantum stochastic differential equations and the master equation. Phys Rev A 31(6):3761–3774. https://doi.org/10.1103/PhysRevA.31.3761
22. Gardiner C, Zoller P (2004) Quantum noise: a handbook of Markovian and Non-Markovian quantum stochastic methods with applications to quantum optics, ser. Springer series in synergetics. Springer, Berlin, Heidelberg
23. Allahverdyan AE, Balian R, Nieuwenhuizen TM (2004) Maximal work extraction from finite quantum systems. Europhys Lett 67(4):565–571. https://doi.org/10.1209/epl/i2004-10101-2
24. Scovil HED, Schulz-DuBois EO (1959) Three-level masers as heat engines. Phys Rev Lett 2(6):262–263. https://doi.org/10.1103/physrevlett.2.262
25. Elouard C, Herrera-Martí D, Huard B, Aufféves A (2017) Extracting work from quantum measurement in Maxwell's Demon engines. Phys Rev Lett 118(26):260603. https://doi.org/10.1103/PhysRevLett.118.260603
26. Cottet N, Jezouin S, Bretheau L, Campagne-Ibarcq P, Ficheux Q, Anders J, Aufféves A, Azouit R, Rouchon P, Huard B (2017) Observing a quantum Maxwell demon at work. Proc Natl Acad Sci USA 114(29):7561–7564. https://doi.org/10.1073/pnas.1704827114
27. Binder F, Correa LA, Gogolin C, Anders J, Adesso G (eds) (2018) Thermodynamics in the quantum regime: fundamental aspects and new directions, ser. Fundamental theories of physics. Springer, Cham
28. Binder FC, Vinjanampathy S, Modi K, Goold J (2015) Quantacell: powerful charging of quantum batteries. New J Phys 17(7):075015. https://doi.org/10.1088/1367-2630/17/7/075015
29. Campaioli F, Pollock FA, Binder FC, Céleri L, Goold J, Vinjanampathy S, Modi K (2017) Enhancing the charging power of quantum batteries. Phys Rev Lett 118(15):150601. https://doi.org/10.1103/PhysRevLett.118.150601
30. Ferraro D, Campisi M, Andolina GM, Pellegrini V, Polini M (2018) High-power collective charging of a solid-state quantum battery. Phys Rev Lett 120(11):117702. https://doi.org/10.1103/PhysRevLett.120.117702
31. Elouard C, Herrera-Martí D, Esposito M, Aufféves A (2020) Thermodynamics of optical bloch equations. arXiv: 2001.08033
32. Carmichael HJ (1999) Statistical methods in quantum optics 1: master equations and Fokker-Planck equations, ser. Theoretical and mathematical physics. Springer, Berlin, Heidelberg
33. Team Commands, Inria Saclay (2017) BOCOP: an open source toolbox for optimal control
34. Valente D, Li Y, Poizat JP, Gérard JM, Kwek LC, Santos MF, Aufféves A (2012) Optimal irreversible stimulated emission. New J Phys 14(8):083029. https://doi.org/10.1088/1367-2630/14/8/083029

Chapter 7
Conclusion

The concept of work is essential to extend thermodynamics to quantum systems and its measurement is a key to the understanding and the experimental exploration of energy exchanges at the quantum scale. However, work measurement is particularly challenging due to the specificities of the quantum world, especially because measurement contributes to energy and entropy exchanges. In this thesis, we have proposed to measure work in situ, directly inside a quantum battery. For this purpose, we have studied two particularly promising platforms to explore the thermodynamics of a qubit: hybrid optomechanical systems and one-dimensional atoms.

First, we have evidenced that in hybrid optomechanical systems, the mechanical resonator plays the role of a "dispersive" battery, providing work to the qubit by modulating its frequency. We have shown that this work can be identified with measurable mechanical energy variations. Furthermore, this method still holds at the quantum trajectory level, allowing to access work fluctuations and therefore to probe fluctuation theorems in an open quantum system. We have demonstrated that shining a detuned laser on the qubit enables optomechanical energy conversion. The direction of this conversion is determined by the sign of the detuning. With a blue detuning, a coherent phonon state can be built starting from thermal noise and the mechanism exhibits laser-like signatures. Reciprocally, with a red detuning, the average phonon number in the resonator can be decreased below the thermal number with a mechanism reminiscent of evaporative cooling. Therefore hybrid optomechanical systems can be seen autonomous thermal machines.

In a second step, we have investigated one-dimensional atoms, consisting of a qubit embedded in a waveguide. We have shown that the modes of the waveguide resonant with the qubit's transition can be seen as a battery allowing both to drive the qubit and to store work, which corresponds here to the coherent fraction of the light emitted by the qubit. Unlike for hybrid optomechanical systems, this battery is resonant, enabling the exploration of the role of coherence in heat-to-work conversion. Exploiting the advanced toolbox of quantum optics, we have suggested to engineer a bath that

J. Monsel, *Quantum Thermodynamics and Optomechanics*, Springer Theses,
https://doi.org/10.1007/978-3-030-54971-8_7

prepares the qubit in an arbitrary superposition of energy eigenstates and, thus, is a source of both energy and coherence. We have evidenced that a two-stroke quantum engine extracting work from a single, non-thermal, bath can be made by combining these engineered bath and battery. The extracted work can be directly measured in the battery, using homodyning or heterodyning techniques. We have finally demonstrated that in the stimulated regime of driving, quantum coherence boosts power extraction, while in the spontaneous regime, it allows to control the engine's yield, acting as a genuinely quantum resource.

The two platforms studied in this thesis open new perspectives for quantum thermodynamics. First, all the numerical simulations in this thesis have been made using experimentally realistic parameters, showing that our proposals are within reach of state-of-the-art experimental devices, allowing for instance to check Jarzynski equality for a quantum open system.

As hinted in Chap. 4, the quantum trajectory picture used to define thermodynamic quantities allows to go beyond the regime of validity of Lindblad master equations, broadening the scope of quantum thermodynamics. Especially, trajectory-based approach have been successfully used to describe non-Markovian processes [1] for which new thermodynamic behaviors are expected [2, 3], for instance due to memory effects [4].

A straightforward follow-up of Chap. 5 would be to investigate further the energy conversion with the hybrid optomechanical system, especially the red-detuned regime. We have evidenced that in this regime the mechanical energy is reduced but we have not shown that the resonator is cooled down in the thermodynamic sense. So the next step would be to check whether the resonator thermalizes and to study the entropy during the process. In addition, a more detailed modeling of the interaction between the qubit and the laser would allow to determine the limit of the mechanical energy reduction. Besides, the laser, used as a hot bath, is a typical example of colored bath. This kind of baths are frequently used in quantum optics and it would be interesting to explore their thermodynamics and derive fluctuation theorems in this context.

As we saw in Chap. 6, bath engineering opens new perspectives in quantum thermodynamics, allowing to explore negative effective temperatures [5] or cases were the bath provides coherence to the system [6]. But this make the definition of thermodynamic quantities like heat less obvious. It would be interesting to analyze more in detail the nature of the energy provided by the bath to the qubit, which requires to study the energy flows inside the specific setup used to engineer the bath, as will be done in Ref. [7]. It would therefore be useful to develop a more generic approach to study engineered bath and evaluate the cost of this engineering.

One of the key features of the two studied platforms is that the battery is quantum and therefore fully integrated in our quantum description of the system. This is a clear departure with respect to the state of the art, where most proposals and realizations are based on classical external operators as work sources. The way we identified work, in the absence of explicit time dependence in the global Hamiltonian, could be generalized to more platforms, paving the way for the thermodynamic analysis of experimental devices whose thermodynamic potential have been overlooked so far.

Our thermodynamic analysis of two specific platforms could be extended to most experimental achievements in optomechanics and solid state quantum optics, providing new insights on energy and information exchanges. This is especially appealing for quantum information, to evaluate the energetic footprint of quantum computing and to determine the resources necessary to generate and maintain either coherence or entanglement. This is all the more relevant in view of the proposals of thermal machines [8] and quantum engines [5] to generate steady state entanglement. A thermodynamic analysis also provides interesting insights on error correction [9]. Besides, entanglement is a resource in quantum information and therefore as been considered as a fuel for quantum engines [10–12]. In the future, it will be interesting to investigate further the relationship between entanglement and the amount of extractable work, namely the ergotropy.

We have proposed two methods to directly measure work inside a quantum battery. The first one allows to access work fluctuations, but in a situation where there is no coherence in the energy eigenbasis of the qubit. In the second one, the battery is resonant, allowing to explore the impact of quantum coherence on work extraction, however we only have access to average energy exchanges. A particularly interesting extension of this thesis would be to identify a platform and protocol allowing the direct measurement of work fluctuations in a genuinely quantum situation where a battery coherently drives a quantum open system into coherent superpositions. In this way, it would be possible to measure entropy production and energetic fluctuations of quantum nature [13, 14], related to the erasure of quantum coherences [15, 16]. Relating measurable work fluctuations to quantum entropy production, would open a new chapter in the study of quantum fluctuation theorems.

References

1. Breuer H-P (2004) Genuine quantum trajectories for non-Markovian processes. Phys Rev A 70(1):012106. https://doi.org/10.1103/PhysRevA.70.012106
2. Whitney RS (2018) Non-Markovian quantum thermodynamics: laws and fluctuation theorems. Phys Rev B 98(8):085415. https://doi.org/10.1103/PhysRevB.98.085415
3. Schmidt R, Carusela MF, Pekola JP, Suomela S, Ankerhold J (2015) Work and heat for two-level systems in dissipative environments: strong driving and non-Markovian dynamics. Phys Rev B 91(22): 224303. https://doi.org/10.1103/PhysRevB.91.224303
4. Bylicka B, Tukiainen M, Chruścński D, Piilo J, Maniscalco S (2016) Thermodynamic power of non-Markovianity. Sci Rep 6:27989. https://doi.org/10.1038/srep27989
5. Tacchino F, Auffèves A, Santos MF, Gerace D (2018) Steady state entanglement beyond thermal limits. Phys Rev Lett 120(6):063604. https://doi.org/10.1103/PhysRevLett.120.063604
6. Harrington PM, Naghiloo M, Tan D, Murch KW (2019) Bath engineering of a fluorescing artificial atom with a photonic crystal. Phys Rev A 99(5):052126. https://doi.org/10.1103/PhysRevA.99.052126
7. Elouard C, Herrera-Martí D, Esposito M, Auffèves A (2020) Thermodynamics of optical bloch equations. arXiv: 2001.08033
8. Brask JB, Haack G, Brunner N, Huber M (2015) Autonomous quantum thermal machine for generating steady-state entanglement. New J Phys 17(11):113029. https://doi.org/10.1088/1367-2630/17/11/113029

9. Fellous-Asiani M et al. Energetic limitations of fault-tolerant quantum computation (in preparation)
10. Zhang T, Liu W-T, Chen P-X, Li C-Z (2007) Four-level entangled quantum heat engines. Phys Rev A 75(6):062102. https://doi.org/10.1103/PhysRevA.75.062102
11. Wang H, Liu S, He J (2009) Thermal entanglement in two-atom cavity QED and the entangled quantum Otto engine. Phys Rev E 79(4):041113. https://doi.org/10.1103/PhysRevE.79.041113
12. Alicki R, Fannes M (2013) Entanglement boost for extractable work from ensembles of quantum batteries. Phys Rev E 87(4):042123. https://doi.org/10.1103/PhysRevE.87.042123
13. Elouard C, Herrera-Martí DA, Clusel M, Auffèves A (2017) The role of quantum measurement in stochastic thermodynamics. Npj Quantum Inf 3(1):9. https://doi.org/10.1038/s41534-017-0008-4
14. Elouard C, Herrera-Martí D, Huard B, Auffèves A (2017) Extracting work from quantum measurement in Maxwell's Demon engines. Phys Rev Lett 118(26):260603. https://doi.org/10.1103/PhysRevLett.118.260603
15. Santos JP, Céleri LC, Landi GT, Paternostro M (2019) The role of quantum coherence in non-equilibrium entropy production. Npj Quantum Inf 5(1). https://doi.org/10.1038/s41534-019-0138-y
16. Francica G, Goold J, Plastina F (2019) Role of coherence in the nonequilibrium thermodynamics of quantum systems. Phys Rev E 99(4):042105. https://doi.org/10.1103/PhysRevE.99.042105

Appendix A
Numerical Simulations

A.1 Numerical Simulations in Chap. 4

The numerical results presented in Chap. 4 were obtained using the jump and no-jump probabilities given by Eq. (4.8) to sample the ensemble of possible direct trajectories. I used the following algorithm to obtain the stochastic trajectories:

1. At $t = t_0$, randomly draw the qubit's state $|\epsilon_0\rangle$ using the equilibrium probability distribution (4.2) while the MO is in state $|\beta_0\rangle$.
2. While $t_{k\text{-th jump}} < t_N$, where $t_{k\text{-th jump}}$ is the time of the k-th jump ($t_{0\text{-th jump}} = t_0$):
   (1) At $t = t_{k\text{-th jump}}$, randomly draw a number $r \in [0, 1]$ with a uniform distribution.
   (2) Integrate, using a Riemann sum, the probability that a jump occurs until the cumulative sum reaches r:

$$r = \int_{t_{k\text{-th jump}}}^{t_{(k+1)\text{-th jump}}} P_{\text{no-jump}}(u)\gamma_{\text{jump}}(u)\mathrm{d}u. \tag{A.1}$$

   The jump rate at time u, $\gamma_{\text{jump}}(u) = \gamma(\bar{n}_{\omega(\beta(u))} + \delta_{\epsilon(t_{k\text{-th jump}}),e})$, is obtained from the Kraus operators (4.4). It depends on the state of the qubit after the k-th jump and the state of the MO at time u. $P_{\text{no-jump}}(u)$ denotes the probability that no jump occurred between $t_{k\text{-th jump}}$ and u, therefore $P_{\text{no-jump}}(u + \mathrm{d}u) = P_{\text{no-jump}}(u)(1 - \gamma_{\text{jump}}(u)\mathrm{d}u)$. The time $t_{(k+1)\text{-th jump}}$ such that the integral equals r is the time of the next jump. The evolution of the hybrid system's state between jumps is governed by the effective Hamiltonian H_{eff} (Eq. (4.5)), therefore the qubit's state remains unchanged, $\epsilon(u) = \epsilon(t_{k\text{-th jump}})$, while the mechanical state evolves with $H_{\text{m}}^{\epsilon(t_{k\text{-th jump}})}$ (Eq. (3.7)).

I first coded this algorithm in python but then rewrote it in C++ to increase the computation speed.

J. Monsel, *Quantum Thermodynamics and Optomechanics*, Springer Theses,
https://doi.org/10.1007/978-3-030-54971-8

The average value of a quantity $A[\vec{\Sigma}]$ is then approximated by

$$\left\langle A[\vec{\Sigma}]\right\rangle_{\vec{\Sigma}} \simeq \frac{1}{N_{\text{traj}}} \sum_{i=1}^{N_{\text{traj}}} A[\vec{\Sigma}_i], \tag{A.2}$$

where $N_{\text{traj}} = 5 \times 10^6$ is the number of numerically generated trajectories and $\vec{\Sigma}_i$ denotes the i-th trajectory.

The reduced entropy production $\sigma[\vec{\Sigma}]$ used in Fig. 4.2 was calculated with the expression (4.30), using the numerically generated values of β_0 and $\beta_\Sigma(t_N)$ in the trajectory $\vec{\Sigma}$. One value of $\beta_\Sigma(t_N)$ can be generated by a single direct trajectory $\vec{\Sigma}$: Below we use the equality $p_{\text{m}}[\beta_\Sigma(t_N)] = P[\vec{\Sigma}]$. Using the expression (4.16) of the probability of the reversed trajectory, the average entropy production becomes:

$$\begin{aligned}
\left\langle \Delta_{\text{i}} s[\vec{\Sigma}]\right\rangle_{\vec{\Sigma}} &= \left\langle \log(\frac{P[\vec{\Sigma}]}{\tilde{P}[\overleftarrow{\Sigma}]})\right\rangle_{\vec{\Sigma}} \\
&= \left\langle -\log\left(p^{\infty}_{\beta_\Sigma(t_N)}[\epsilon_\Sigma(t_N)] \prod_{n=1}^{N} \tilde{P}[\Psi_\Sigma(t_{n-1})|\Psi_\Sigma(t_n)]\right)\right\rangle_{\vec{\Sigma}} \\
&\simeq \frac{-1}{N_{\text{traj}}} \sum_{i=1}^{N_{\text{traj}}} \log\left(p^{\infty}_{\beta^i_\Sigma(t_N)}[\epsilon^i(t_N)] \prod_{n=1}^{N} \tilde{P}[\Psi^i_\Sigma(t_{n-1})|\Psi^i_\Sigma(t_n)]\right),
\end{aligned}$$

and,

$$\begin{aligned}
\sum_{\vec{\Sigma}} \tilde{P}[\overleftarrow{\Sigma}] &= \sum_{\vec{\Sigma}} p^{\infty}_{\beta_\Sigma(t_N)}[\epsilon_\Sigma(t_N)] p_{\text{m}}[\beta_\Sigma(t_N)] \prod_{n=1}^{N} \tilde{P}[\Psi_\Sigma(t_{n-1})|\Psi_\Sigma(t_n)] \\
&= \left\langle p^{\infty}_{\beta_\Sigma(t_N)}[\epsilon_\Sigma(t_N)] \prod_{n=1}^{N} \tilde{P}[\Psi_\Sigma(t_{n-1})|\Psi_\Sigma(t_n)]\right\rangle_{\vec{\Sigma}} \\
&\simeq \frac{1}{N_{\text{traj}}} \sum_{i=1}^{N_{\text{traj}}} p^{\infty}_{\beta^i_\Sigma(t_N)}[\epsilon^i(t_N)] \prod_{n=1}^{N} \tilde{P}[\Psi^i_\Sigma(t_{n-1})|\Psi^i_\Sigma(t_n)].
\end{aligned}$$

To obtain Fig. 4.3, we considered that the preparation of the initial MO state was not perfect. So instead of starting from exactly $|\beta_0\rangle$, the MO trajectories start from $|\beta_\Sigma(t_0)\rangle$ with the $\beta_\Sigma(t_0)$ uniformly distributed in a square of width $2\delta\beta$, centered on β_0. Similarly, the measuring apparatus has a finite precision, modeled by a grid of cell width $2\delta\beta$ in the phase plane $(\Re\beta_{\text{f}}, \Im\beta_{\text{f}})$. Instead of obtaining the exact value of $\beta_\Sigma(t_N)$, we get $\beta^{\text{M}}_\Sigma(t_N)$, namely the center of the grid cell in which $\beta_\Sigma(t_N)$ is. The value used to compute the thermodynamical quantities are not the exact $\beta_\Sigma(t_0)$ and $\beta_\Sigma(t_N)$ but $\beta^{\text{M}}_0 = \beta_0$ and $\beta^{\text{M}}_\Sigma(t_N)$.

A.2 Numerical Simulations in Chap. 5

The numerical trajectories plotted in Figs. 5.5 and 5.6 in Chap. 5 were obtained with the following algorithm, obtained from the unraveling described in Sect. 5.2.1.3:

1. The qubit is initialized in $|g\rangle$ and the MO's state $\beta_0 = r_x + \mathrm{i}r_y$ is drawn from a thermal distribution. More precisely, r_x and r_y are two random numbers drawn from independent Gaussian distributions of zero mean and standard deviation $\sqrt{N_{\mathrm{th}}/2}$.
2. For each time step n between 1 and N:

   (1) At $t = t_n$, draw the random numbers r_x and r_y from independent Gaussian distributions of zero mean and standard deviation $\sqrt{\Gamma \Delta t N_{\mathrm{th}}/2}$. $r_x + \mathrm{i}r_y$ corresponds to the Wiener increment $\sqrt{\Gamma N_{\mathrm{th}}}\mathrm{d}\xi_+(t_n)$ from Eq. (5.42).
   (2) Compute $\beta_\Sigma(t_n)$ using Eq. (5.42):

$$\beta_\Sigma(t_n) = \left(\beta_\Sigma(t_{n-1}) + \frac{g_{\mathrm{m}}}{\Omega}\delta_{\epsilon_\Sigma(t_{n-1}),e}\right)\mathrm{e}^{-(\mathrm{i}\Omega+\frac{\Gamma}{2})\mathrm{d}t} - \frac{g_{\mathrm{m}}}{\Omega}\delta_{\epsilon_\Sigma(t_{n-1}),e} + r_x + \mathrm{i}r_y. \tag{A.3}$$

   (3) Check whether the qubit interacts with the laser, i.e. whether $\omega(\beta_\Sigma(t))$ crossed the frequency ω_{L} during the n-th time step:

   - If the qubit interacts with the laser, draw the state of the qubit after the interaction (probability θ to be excited). Then, if the qubit is in the excited state, randomly draw the time of the spontaneous emission from an exponential distribution of decay rate γ.
   - Otherwise, check whether the spontaneous emission occurred during this time step and change the qubit's state accordingly.

I first coded this algorithm in python, then rewrote it in C++ for speed reasons.

Curriculum Vitae

Juliette Monsel

Researcher in theoretical physics

Göteborg, Sweden
monsel@chalmers.se
www.linkedin.com/in/juliette-monsel

Research interests: stochastic thermodynamics, quantum optics, optomechanics and electronic transport.

Education

2019 **Ph.D. in theoretical physics**, *Université Grenoble Alpes (UGA),* Grenoble, France.

2016 **M.Sc.**, *École Normale Supérieure de Lyon,* France, Major: Physics, Mention: highest honors.

2014 **B.Sc.**, *École Normale Supérieure de Lyon,* France, Major: Physics, Mention: highest honors.

Experience

Research

2020–Current	**Postdoctoral researcher**, *Chalmers University of Technology*, Göteborg, Sweden, in the group of Janine Splettstoesser.
2019–2020 (4 months)	**Postdoctoral researcher**, *Institut Néel CNRS/UGA,* Grenoble, France, in the group of Alexia Auffèves.
2016–2019 (3 years)	**Ph.D. student**, *Institut Néel CNRS/UGA,* Grenoble, France, supervisor: Alexia Auffèves. Quantum thermodynamics and optomechanics.

J. Monsel, *Quantum Thermodynamics and Optomechanics*, Springer Theses,
https://doi.org/10.1007/978-3-030-54971-8

2016 (4 months) **Master thesis**, *Institut Néel,* Grenoble, France, supervisor: Alexia Auffèves.
Fluctuation theorems in a hybrid optomechanical system.

2015 (3 months) **Internship**, *Institut Néel,* Grenoble, France, supervisor: Alexia Auffèves.
Hybrid optomechanical system in the ultra-strong coupling regime.

2014 (2 months) **Bachelor thesis**, *Institut Lumière Matière,* Lyon, France, supervisor: Julien Laverdant.
Experimental control of polarization with a spatial light modulator.

Teaching

2017–2018 (129 hours) **Teaching Assistant**, *Université Grenoble Alpes,* Grenoble, France.
Newtonian mechanics for first year undergraduates.

- Supervised students during tutorials and practical work
- Graded examinations and practical work reports

2013–2014 (7 months) **Tutor for homework assistance**, *Trait d'Union program,* Villeurbanne, France.
Homework assistance program for students from high schools in disadvantaged areas (two hours a week).

Volunteer Work

2020–Current **Cykelköket**, Göteborg, Sweden.
Helped people repair their bikes at the "Bike kitchen", an open Do-It-Yourself bicycle workshop.

2017–2020 **uN p'Tit véLo dAnS La Tête**, Grenoble, France.
Associative self-repair workshop aiming at teaching bicycle mechanics and promoting bike riding.

- Learned bicycle mechanics by dismantling and repairing bikes for the association
- Explained to members of the association how to repair their bikes
- Took part in meetings and helped organize events as a member of the board from September 2018 to February 2020

Skills

Languages **French**: native **English**: fluent **Italian**: average
Computer **Programming**: Python, LaTeX, Git **Softwares**: Matlab, Spyder

Publications

2020 J. Monsel, M. Fellous-Asiani, B. Huard, and A. Auffèves. "The Energetic Cost of Work Extraction". *Physical Review Letters*, **124**, 130601.

2019 J. Monsel, M. F. Asiani, B. Huard, and A. Auffèves. "A coherent quantum engine based on bath and battery engineering". In *Rochester Conference on Coherence and Quantum Optics (CQO-11)*, p. W2A.2. Optical Society of America, 2019.

2018 J. Monsel, C. Elouard, and A. Auffèves. "An autonomous quantum machine to measure the thermodynamic arrow of time". *npj Quantum Information*, **4**, 59.

Conferences and Summer Schools

2019 **Fourth Annual Meeting of the GDR MecaQ**, Palaiseau, France.
Contributed talk: "An autonomous optomechanical energy converter".

2019 **Quantum ThermoDynamics Conference**, Espoo, Finland.
Contributed talk: "An autonomous quantum machine to measure the thermodynamic arrow of time".

2019 **II Workshop on Quantum Information and Thermodynamics**, Natal, Brazil.
Contributed talk: "An autonomous quantum machine to measure the thermodynamic arrow of time".

2019 **Workshop on Quantum Networks and Non-equilibrium Systems**, Obergurgl, Austria.
Invited talk: "An autonomous quantum machine to measure the thermodynamic arrow of time".

2018 **Condensed matter days (JMC)**, Grenoble, France.
Invited talk: "Energy conversion in a hybrid optomechanical system: Laser-like behavior and cooling".

2018 **Thermodynamics of quantum systems: Measurement, engines, and control**, *KITP*, Santa Barbara, California, United States.
Took part in the program during one month including the one week conference and interacted with scientists from my field.

2017 **Quantum Engineering, from Fundamental Aspects to Applications (GDR IQFA) - 8th colloquium**, Nice, France.
Contributed talk: "Fluctuation theorems in a hybrid optomechanical system".

2017 **VI Quantum Information School and Workshop**, Paraty, Brazil.
Contributed talk and poster: "Measuring the arrow of time in a hybrid optomechanical system".

2017 **Congress of the French Physical Society**, Orsay, France.
Invited talk: "Thermodynamics and hybrid optomechanical system".

2017 **Fifth Quantum Thermodynamics Conference**, Oxford, United Kingdom.
Contributed poster: "Measuring the arrow of time in a hybrid optomechanical system".